The Social
Primates

**Animal Behavior Series Under the Editorship
of Vincent G. Dethier and H. Philip Zeigler**

The Social Primates

Paul E. Simonds

University of Oregon

HARPER & ROW, PUBLISHERS
New York, Evanston, San Francisco, London

Sponsoring Editor: George A. Middendorf
Project Editor: David Nickol
Designer: June Negrycz
Production Supervisor: Robert A. Pirrung

Library of Congress Cataloging in Publication Data

Simonds, Paul E   1932
    The social primates.

    (Animal behavior series)
    1. Primates—Behavior.   2. Social behavior in animals.
I. Title.   II. Series.
QL737.P9S49   599'.8'045   73-13300
ISBN 0-06-046165-9

Dedicated to Theodore D. McCown

Contents

CHAPTER 11   The Adaptive Nature of Primate Society   223

Plates

Acknowledgments

I wish to acknowledge those whose advice,
teaching, and support led to this work in
a multitude of ways: S. L. Washburn,
whose ideas I have internalized to such
an extent that I can no longer tell where
his leave off and mine begin; Margaret
Barrs, whose editing clarified the end
product; the National Institute of Mental
Health, whose support enabled me to
conduct fieldwork; the scientists at the
Japan Monkey Center, who first introduced
me to fieldwork; the Government of India,
which allowed me to work in the forests
and countryside of south India; my wife,
for her patience; and first, last, and always,
Theodore D. McCown, whose teaching
inspired my devotion to anthropology and
stimulated what capacities of
mind I possess.

The Social
Primates

Introduction

Among the many major reasons for studying primates, the following are probably the most important.

As man's most immediate relatives, nonhuman primates, especially the higher species such as monkeys and apes, share a common biological heritage with him. Because of this heritage, they depend on similar mechanisms for acquiring stimuli from their environment and react to such stimuli much as man himself does. Possessing stereoscopic color vision, higher primates rely on the sense of smell at about the same low level as man. In contrast to most other mammals, they use their hands a great deal to manipulate and have, therefore, a sense of touch and a related cortical control of the hand that is comparable to man's.

As laboratory subjects, primates enable us to conduct experiments not generally possible with human beings because of individual privacy and moral strictures. Moreover, since most nonhuman primates are small, have a shorter life span, and grow and develop at a faster pace than man, experimental results are achieved more quickly, cheaply, and easily than if man were the experimental animal.

As highly advanced social mammals resembling man in many ways, nonhuman primates also provide invaluable information about the function and structure of society. Much of this book is concerned precisely with this aspect of the nonhuman primate's contribution to better human living.

PAST MISCONCEPTIONS

An astounding number of false notions about monkeys and apes have been used to justify one or another theory about the development of human behavior and institutions. Unfortunately, these misconceptions, which have developed over long periods of time and have become well established, have tended to obscure the discussion of process and history in the social sciences. However, more recent studies of primates have done much to disabuse us of these scientifically unsound notions about primate behavior.

From the beginning of recorded history, man has been aware of the primates. The ancient Egyptians incorporated baboons into their religion as sacred animals and pictorially represented other monkeys, probably African guenons. Their art forms depict several different primate species, with some of which, at least, they were acquainted. Totally absent, however, is any indication that they recognized the phylogenetic relationship between man and the other primates.

Later evidence indicates that the Greeks and Romans recognized the resemblance between monkeys and man and speculated about their relationship; whether they delved any further into the matter is not known. Some philosophical speculation about the relationship filtered into Renaissance writings and later records. But since the observations were confined to rare captive animals far from their native element and came from areas that could have had no first-hand knowledge of wild primates, their value and interest are historical rather than scientific.

From the seventeenth century almost to the present, the discussion of primate behavior tended to anthropomorphize (that is, to assign human motives to) their behavior or to resort to anecdotal reports based on exaggerated tales of natives or on exceedingly brief observations by the narrator, made perhaps from the safe vantage of a hotel veranda over a long drink!

With the age of exploration came the influx of thousands of Europeans into those parts of the world where primates and "primitive" people lived. In the minds of many of these travelers, there was little difference between the "naked savage" and the "hairy ape." Local legends and primitive concepts, which endowed monkeys and apes with something of human status, were accepted by the Europeans at face value. To credulous Europeans taking it all in for the first time, the orang utan or "man of the woods" might have seemed, on fleeting observation, only a little more strange than the orang laut or human "men of the sea" who peopled the Indonesian archipelago. The legends of the armies of Hanuman, who supported Rama in his quest for his abducted bride, must have given the monkeys of India a near-human status in the minds of some of those early adventurers. Their reactions to the myriad strange and wondrous sights are not too hard to understand, nor is the credulity that led them to assume that the primates they saw in the forests and temples were more marvelous than in reality.

To further erase the thin line between fact and fiction, the adventurers on their return home embellished their tales for the titillation of their audiences. Brief observations were mixed with local legends to concoct a tale well salted with imagination and peppered with exaggeration. The results have so obscured the truth that even today it is impossible to know or conjecture what primates were being described in those reports. In some cases, the characteristics of several different species were assigned to one. In others, human pygmies were described as apes and apes as humans, though of a rather strange aspect.

SCIENTIFIC MISCONCEPTIONS

Malinowski, among other anthropologists earlier in this century, had a number of misconceptions about primate sexual behavior that were the source of many other false notions. He accepted as fact that monkeys have a rutting season; that is, that their sexual behavior is based primarily on an automatic physiological response in which a releasing mechanism triggers sexual behavior. His entire discussion on the origin of human sexuality was based on this assumption. But as a matter of fact, primates have an extremely varied and diverse sexual behavior that is controlled not by instinct or reflex responses but rather by learning. Male and female primates do not indulge in sexual behavior because their physiology has produced conditions of rut and estrus but because they have learned the proper responses in specific situations. Under certain circumstances, a female in estrus will refuse to mate with certain males in her society. Again, a female will not mate or be mated with unless she (and her partner) have learned the proper techniques and the social proprieties involved through a long period of socialization. Moreover, even the male primate will not respond to an estrous female presenting to him sexually unless the social setting and various other circumstances are correct.

Recent studies of animal behavior, particularly of primate behavior, show that much of it is learned. The variety of behavior among different troops of monkeys of the same species can only be the result of learning. Among primates and many other animals, the genetics of behavior tends not so much to determine behavior patterns as to set limits and define directions; in other words, it is a genetics of tendencies. This is, of course, contrary to Malinowski (1960), who wrote: "But while collective behavior in animals is due to innate equipment, in man it is always a gradually built up habit." In other words, animals rely on instinct, whereas man relies on learning.

As a logical deduction from these theories, Malinowski also believed that to describe the behavior patterns of one group of animals of a single species was to describe the typical behavior of that species and hence that all other groups could be expected to display the same behavior. In other words, he believed that since behavioral patterns within a single species do not vary, the behavior of one group is the key to a complete understanding of the entire species. But long and careful studies by Japanese primatologists of native Japanese macaques have conclusively shown that some group traditions are learned and thus vary from group to group.

Even as late as 1964, we find erroneous reports about the nature of primate behavior. In his book *Primitive Social Organization,* Service writes, "The sexual behavior of a species is biological and a constant for the species." But field studies have shown that the sexual behavior of a single species of monkey varies remarkably not only among troops of the same species but even within a single troop. Other, even broader generalizations about the behavior of monkeys are not applicable even within a single species and therefore certainly cannot be valid for all monkeys and apes, for prosimii and anthropoidea.

Since these recent generalizations by social scientists are being used to trace the behavior patterns from which human society and behavior evolved, they must be carefully checked against factual data obtained from field studies of the animals concerned.

EVOLUTION OF HUMAN BEHAVIOR

In correcting the past and current misconceptions about primate behavior, we must concern ourselves with exposing and correcting the consequent misconceptions about the evolution of human behavior. We can no longer generalize with assurance about the primeval promiscuous horde once thought to have been the source of man's complex kinship system. Nor are we justified in assuming that a harem organization preceded the more complex systems of human social organization. Both conditions exist in wild primate societies. Either the horde or the harem or some as yet undescribed pattern of social organization may have characterized some of our own ancestors. Although it is unlikely, the family organization of the gibbon could be an alternative to the two systems just mentioned.

In the long run, careful comparisons of major groups of primates and of species within those groups should yield some valid generalizations that can be used as working hypotheses to aid us in understanding the form of primate behavior from which human behavior has evolved. Ten to twenty million years ago hominid or hominoid group organization probably involved bands similar to those found today among gorillas, chimpanzees, and baboons. They may not have been as inflexible as baboon troops or as changeable as chimpanzee bands. They may have had features that are absent from chimpanzee, baboon, and gorilla groups of today. But we can reasonably assume that the groups included individuals who knew their own status in the group, understood the relationships between the other individuals in the group, and tended to operate socially within the system of relationships that bound them together. We can further assume that the groups were relatively permanent. These assumptions are reasonable since the features are found in all higher primate societies now known.

The study of primate society in some respects makes human society stand out starkly against the background of other primates. Our communication system obviously differs from that of other primates. Language allows us a much richer system of learning and communicating information and thus makes possible a greater elaboration of behavior. The complicated technology that characterizes even nonindustrial societies contrasts strongly with the almost total lack of tool-using among other primates.

We can, then, reconstruct in general outline what our ancestral behavior was like and then demonstrate the major changes that have brought us to our present human condition. A great deal of caution must be exercised in this process, however. Behavior is an extremely changeable factor. Unlike cusp patterns on molar teeth, which change very slowly over long periods of time and require generations of gene recombinations, mutation, and selection, a new pattern of behavior can be introduced within the life span of one individual and can be adopted by all the members of that particular society while neighboring societies remain unchanged. This is as true of monkeys as it is of man. It is quite conceivable that in several respects the line of primates leading directly to man had a very different kind of behavior from that of other primates and that, in fact, they must have differed in some highly significant ways for the whole process of evolution to have forced them off into a new direction. It is also conceivable that the other primates who shared behavior patterns with man's direct ancestors were eliminated by competition with the more successful hominids and therefore are no longer represented in today's living primates. It must further be recognized that the other living primates, which have had an equal span of time in which to evolve their own particular behavioral adaptations, may have diverged considerably from the behavior of our common ancestors.

THE FIELD STUDIES

The current division of primate studies into field and laboratory studies is not necessarily beneficial. The result is that most primatologists specialize in either one or the other and often do not understand the problems of their colleagues working in the other field. My own work has been in field studies and my biases will be obvious to those who have had primarily laboratory experience.

It is important at the outset to understand the differences in technique involved in field and laboratory work. Those who work with wild, free-ranging primates must meet them on their own terms (or not at all!) and must make a series of concessions for the privilege of watching and recording their behavior. At best, only a minimal control can be exerted over the free-ranging animal; on some occasions and under certain circumstances, even that minimum is impossible.

The first condition for observing primates in the wild is so obvious that it almost seems redundant to mention it. That condition is proximity and proximity involves trust. In some cases, the subjects of the study live in areas where close contact with man is common; here there need be a conditioning period of only a few days to a few weeks. Under these conditions, accurate and usable data can be collected even on the first day of the study. In other cases, particularly where the animals have been hunted, long periods of patient maneuvering must be endured before the primates are won over.

My own experience with bonnet macaques in southern India is a case in point. The roadside troops became accustomed to my presence in less than a month, treating me as an insignificant part of their environment and presenting a complete (or nearly complete) repertoire of their behavior. However, forest troops, acquainted with hunters, refused to accept my presence even after numerous contacts spread over six months; the only behavior pattern I could observe was flight.

However, the Baroness Van Lawick-Goodall has demonstrated the power of patience. She spent nearly two years habituating the chimpanzees of the Gombe Stream Reserve to her presence before she could attempt systematic behavioral observations. Van Lawick-Goodall's strategy was to station herself where she could be clearly observed and to avoid any movement or action that could be considered aggressive. Japanese primatologists ordinarily use the stratagem, when possible, of feeding the wild primates they wish to study. However, this technique is not effective with monkeys whose diet consists of fresh buds, shoots, and leaves.

IDENTIFICATION OF INDIVIDUALS

Careful field studies of primates depend upon the identification of individual animals. The observer cannot, for example, say anything about the differences between the sexes in terms of play, grooming, dominance, and sex unless he can identify the males and females. By the same token, he cannot demonstrate dominance hierarchies unless he has first identified the high-ranking and low-ranking animals. To collect meaningful data, the observer must be able to record with assurance that animal α threatened animal β five times this week and that animal β did not threaten animal α.

Several methods are used to identify individual animals. In some situations the members of the group under study can be marked. For example, rhesus macaques, *Macaca mulatta*, maintained on Cayo Santiago, Puerto Rico, have been tattooed in conspicuous places or tagged with metal eartags. Such markings make for relatively easy comparison of data among the various observers.

Unfortunately, this means of marking very often involves the necessity of trapping the animals beforehand. But among many groups of wild primates, trapping to tattoo the animals or to mark them with paint would disrupt the group and cause the animals to be so wary of the observer (even if someone other than the observer did the trapping and marking) that he could not achieve the close contact necessary for observations.

Most studies done in forests or on savannahs rely on the individual characteristics of the monkeys and apes to establish identification. The obvious signs are learned first. The male with two-thirds of his tail missing is recognized and named or numbered in the notes on the first day; the female with most of her left ear torn away is noticed and named on the second day. Other less obvious individual differences are soon recognized, and it is not long before most of the individuals in a troop can be characterized not only by their clearly unmistakable traits but also by the way they walk, sit, or play. In less than a month of continuous observations thirty to fifty animals can easily be recognized in this manner. Small troops of five to fifteen animals can often be identified in a day or two. The hardest individuals to identify are the young, who have not yet been knocked about enough to show life's scars and are growing so fast that their physiognomy changes rapidly.

Once the observer has identified the individuals in the group, he can begin to make reliable observations. His data then become the basis of his analysis of social structure. Subgroupings can be either demonstrated or reliably denied. Status and role systems in the society can be recognized, and changing relationships, which show the dynamics of the society, can be distinguished. The patterns of troop progression and spacing can be plotted. Without the basic identification data, such findings would be impossible.

LONG-TERM OBSERVATIONS

It cannot be emphasized too strongly that the longer the observations, the better the generalizations based upon the data are likely to be. Some patterns of behavior are seasonal or cyclical. Many primates have a mating season during which most copulations take place and a birth season during which most of the infants are born. Thus short observational periods can show only extremely limited aspects of such major behavior systems as the mother–infant relationship.

Other patterns of behavior are rare or occur under special circumstances. For example, the reactions of troop members to the death of one of them might be observed only once in several years of observations, if at all. Predation may affect the troop only once or twice a year. Social relationships may be stable for months or years and be disrupted only when an outsider joins the group or when a subadult male approaches physical and social maturity after six to seven years of growth.

Most modern field studies of primates are planned for at least a year, and a single year's study is considered preliminary to further work. The first year enables the field worker to study a small number of troops fairly intensively and to work out the major outlines of species behavior. Confined to a few troops, often in only one or two areas of the total range, a single year's study cannot possibly include the variations in different ecological settings. An adequately detailed study involves repeated or continuous field work on many different groups in all of the ecological settings in which the species is found. Laboratory work, designed to supplement the field work, can quite profitably be pursued between field trips. Problems that arise in the field can be tested in the laboratory and then restudied in the field with the new knowledge and insight derived from the laboratory.

Japanese primatologists hold the best record for maintaining long-term studies. Many of them work on the same groups, which are kept under continuous observation year after year, sometimes for longer than twenty years. In some troops of Japanese macaques, the life histories of some adults have been recorded from the time of birth. Thus for the first time the social development of individual primates in free-ranging troops has been demonstrated. Only under these conditions can one state with certainty that a certain male is the son of a certain female and compare his status in the troop with that of his mother.

The dynamics of social interaction also require long-term studies to be understood. Changes in troop leadership, for example, may be slight over a period of twenty years, during which a group that had gained control in its youth maintains it until the members die of old age one to two decades later. On the other hand, a change in leadership that is exceedingly slow at one point in the history of a single group may occur suddenly at another time. Observations of one or the other of these incidents but not of both would give an exceedingly lopsided picture of social dynamics.

Long-term events that occur cyclically over periods of years cannot be accurately observed or reported by short-term studies. In the forests of southern India, where their habitat is relatively undisturbed, bonnet macaques, *Macaca radiata,* live primarily in large bamboo clumps. Bamboo dies when it goes to seed and reforestation of large tracts sometimes takes place at cyclical intervals. The species of bamboo in which bonnet macaques live goes to seed every fifteen years and the monkeys may be forced to move to new areas until the young bamboo again supplies the necessary food and shelter. No short study is likely to reveal cyclical changes such as these; as a result of such studies, we are probably abysmally ignorant of all but an infinitesimal fraction of the living cycles of primates.

Droughts, floods, major shifts in ecology, though not generally cyclical, occur again and again over long periods of years, and their effects must be highly significant in the behavior of animals that have to rely heavily on past experience to adjust to new or to repeating situations.

GROUP BEHAVIOR IN THE NATURAL SETTING

Primate behavior did not evolve in isolation. The selective pressures that have shaped primate patterns of social behavior have operated for many millions of years in a rich natural environment where the influence of other animals (predators and otherwise) both vertebrate and invertebrate, plants, and the weather have played interrelated and interdependent roles. Primates have adjusted to this rich environmental setting by developing an elaborate social setting, by no means unique to the primates, in which cooperation between the individuals ensures the survival of the group. Many mammals, birds, and insects, and other forms as well, have developed cohesive social groupings for survival.

The major advantage of field studies over laboratory research lies precisely here: in the natural setting the animals are constantly adjusting to that setting and to ecological vagaries that threaten their daily activities and their survival as a group. Since the society is seen to function within this matrix, the relation of the behavior patterns to the survival and functioning of the group can be worked out. Some of these relations are obvious, such as the survival value of the infant's grasping reflex as the mother jumps from branch to branch with the infant clinging to her belly; others are less so and require long, detailed study.

Only field studies make available for examination the complexities and close relationships between the ecological niche and the characteristics of the individual primate and its functioning within and contribution to society. A single species is often found in several related but somewhat different ecological zones; the society tends to shift its character somewhat from zone to zone. The observer is confronted by a society with a history of untold years of successful adaptations to its credit. The present successful adaptation can be examined from a multitude of aspects, and if the observer is fortunate (or manipulative), he can observe the society in its day-to-day activity as well as its reaction to crises. Long-term studies can probe the processes of social change and the slower shifts of adaptation.

The natural setting varies somewhat from group to group; in some cases a single species occupies several rather diverse ecological niches. Correspondingly the different groups of a single species vary in group composition and patterns of interaction. Hence field studies should not be confined to single groups but should sample the range of behavioral variation from group to group. This variability, which shows the flexibility of the primate response, provides the background for controlled laboratory studies.

Field studies, then, give us the total view of the society successfully operating in its environmental setting, the whole rich spectrum of inter-actions that disclose the evolutionary significance of single acts and complex patterns of activity.

LABORATORY STUDIES

Laboratory studies of primates, by definition, involve a much greater control over the animals and the situations in which they are observed. The animals are removed from their natural habitat and housed under artificial conditions. Their freedom to move about is severely restricted and their contact with other members of their own species and other animals is reduced or eliminated. They are fed at certain times during the day at the desire of their keepers or the requirements of the experiment. Their sexual and general reproductive behavior is usually restricted and controlled.

Laboratory studies are carried out on single individuals, pairs, or groups of larger size. The tendency in the past has been to emphasize individual or paired animals, partly because they are easier to handle and partly because it is desirable to have a clear understanding of individual behavior before experimenting with group behavior. For this reason, individuals have usually been isolated except for observations in controlled experimental settings. Thus interactions being studied are between individuals that may know one another from previous brief encounters but that have not been able to work out the subtle means of interacting without overt aggression.

More recently, workers like Harlow have refined and extended the study of dyads, keeping the monkeys together for long periods of time to discover the ontogeny of social relationships and the slow development of particular sets of interaction such as mother-infant or peer group relationships.

CONTROL OF VARIABLES

The major advantage of laboratory studies is that the variables can be controlled. In the natural setting, the extremely rich context that is a key factor in developing an understanding of the total adaptation of the species is a drawback in defining such relationships as the mother–infant bond and the extent and limits of its contribution to a viable and complete adult. The variables are far too numerous to be controlled in any satisfactory manner.

In the laboratory, the observer can pare away irrelevant aspects of the environment until he has just the set of variables he wishes to test or to hold constant. Patterns of behavior can be isolated, and the factors that influence them and their effect on other patterns of behavior can be investigated. The chapter on adult–young relationships illustrates the contribution of this approach.

Not only external variables such as outside influences but internal variables as well can be controlled in laboratory situations. The nature of light and sound, the amount of space, the recency of a meal, and many other physical and physiological factors are important and must be regulated either directly or indirectly by holding the overall setting as constant as possible. An important aspect of the laboratory control of variables, and one that is very difficult to control in the wild, is the number of individuals allowed to interact. Much of recent laboratory work deals with the ratios of various categories of animals in the group and the effect of changing ratios. Such factors can be manipulated: one can set up a group with additional males or an overbalance of females. One can arrange play groups composed of one or the other sex, or of both, with many individuals of the same age or of widely different ages. Each situation reveals important facts about the ontogeny of behavior and the role of different categories of individuals within the social setting.

Long-term studies are more and more important in the laboratory as well as in the field. No longer will allowing two or three animals to interact for fifteen minutes on half a dozen occasions answer the complex questions that are being raised. The animals must become accustomed to each other, must work out their own means of interacting smoothly before the experiment can produce results. Working out the interactions is important; but if one is seeking to understand the processes of adaptation, he can apply external pressures only to a society that is adjusted internally. In cases where the reactions of a natural group are to be contrasted with those of an artificially constructed group, it is desirable to capture intact a wild group of the right composition and to bring it into the laboratory. They must, of necessity, be studied for months or even years before the data can be reliable and pertinent.

THE RELATION OF LABORATORY TO FIELD STUDIES

Laboratory and field studies together, by their very differences, contribute to the total picture of primate behavior. Behavior must be seen in its natural context, in relationship with other kinds of behavior, and in the setting in which it evolved if the functional implications of complex behavior patterns are to be understood. On the other hand, unless variables can be controlled, the interplay of one facet of behavior with another in the developing individual or in the developing social situation cannot be observed and interpreted correctly.

Such problems as how much learning and how many inherited behavior patterns are involved in primate communication cannot be solved without controls. In the laboratory, where the individual animal can be isolated from any social contact, one can determine whether the sounds and gestures typical of the species develop under such conditions and whether the animal uses them appropriately, which parts of the communication system are actually inherited and which must be learned in the context of a multifaceted social group.

On the other hand, the laboratory situation cannot show the whole range of meaning that can be transmitted by the communication system. It lacks the multitude of factors that constantly impinge upon the wild primate's consciousness and determine, to a great degree, the response of the individual to a particular stimulus at specific times. Some patterns of communication are never seen or heard in a laboratory situation and can be discovered only under rare circumstances in the wild.

Where does the study of primate behavior begin or end? Field studies raise problems that can only be solved in the laboratory, and solving them under controlled conditions raises new questions that can only be answered in the field. Someone concerned with primate communication observes certain facial gestures in the wild. They are recorded, interaction patterns are worked out between individuals of different rank and status in the society, and a series of expectations are generalized from the observations. However, he must turn to the dissection laboratory to learn which muscles and sets of muscles were coordinated to produce a particular gesture. Once he knows these physical prerequisites, he returns to the field and observes the gestures with a better understanding of their mechanics, able now to analyze the finer variations of each gesture in greater detail. Often ambiguities in the earlier field studies are thus cleared up. But the earlier studies were necessary before he could even discern the importance of facial musculature in communication.

Only by using the two complementary methods of investigation will the student of primate behavior avoid (1) the stagnation of repeating one field study after another, hampered by the inability to control variables effectively or (2) the proliferation of dyadic interrelationship studies or some similar activity without regard for the adaptive qualities of primate behavior or its variability when exposed to many factors at once. The one-method approach leaves many questions unanswered, and unanswered questions do not advance the study of primates.

Chapter 1
Classification

Before the nature and characteristics of any series of objects can be discussed, they must be named. In addition, the relationships between objects must be known, although the relationships may be the subject of the discussion since there is often disagreement about relationships in taxonomic systems. Every system of classification attempts both to name and to state relationships. The study of primate behavior is concerned with an order of mammals, and any systematic comparison of the behavior of the different forms of primates must be based on an understanding of their relationships. The classification presented here, based broadly on Simpson (1945), is abbreviated in many respects. It is an attempt to order the forms of primates whose behavior has been studied in relationship to primates as a whole in such a way that the reader can follow the behavior of closely related forms and can see how they tend to contrast with more distantly related forms. It may also help to show that closely related forms that have adapted to rather different environments have markedly altered their behavior. The best example of such alteration is the behavior of the closely related chimpanzee and man. They are similar in many aspects of their anatomy, biochemistry, and nervous systems, but chimpanzee behavior and human cultural behavior are very different in character and in the way they affect the environment.

TAXONOMIC CATEGORIES—THE SPECIES

In much the same way that the atom is the basic unit in modern chemistry and the gene in modern genetics, the species is the basic unit in evolution and in taxonomy. Just as the atom and the gene can be divided, so the species can be subdivided into populations and ultimately into individuals. It is the only unit in the taxonomic system that is not defined with reference to another taxonomic unit.

The species is the base upon which the higher categories are constructed and constitutes the conceptual whole for the smaller units. The evolutionary unit, the species is the population within which interbreeding takes place. The subspecies and varieties and races can merge again into the species even though they may go on to evolve into separate species. The higher categories (for example, genera, families, orders) are, for the most part, distinct and separate. They cannot merge again into single interbreeding populations or panmictic gene pools. The species, then, is the largest self-contained natural unit.

The interaction between the members of a species differs from that between member species of a genus or member genera of a family. Interbreeding, the formation of new gene pools from preexisting ones, is possible among the members of a single species. In the higher categories, the meeting between two genera or two families as they compete for the same ecological niche is just that, competition rather than cooperation, and the result is generally the extinction of one form rather than a recombination of genes.

Moreover, the species is the basic unit of study for the behavioral scientist. Although he may work with individual social groups, his goal is to determine the modal behavior pattern for the species and the range of variability in species behavior and thereby to establish a basis of comparison of behavior among species. Individual studies of single groups of animals, necessary as a beginning, are limited in their usefulness until enough of them have been conducted to present a clear picture of the usual patterns of behavior and to define the limits for the species.

The gene pool of the species is the common base upon which the variability in behavior is developed. We cannot, of course, assume that all differences in behavior are the result of learned responses to new and different situations, but the species' genetic background probably has many more genes that are common throughout the gene pool than genes that determine different behavior patterns in different environments. Thus, particularly for primates, which rely heavily on learned behavior, most differences in behavior patterns within a species probably result from learned responses rather than from altered gene frequencies. In the future, we may be able to isolate behavior patterns that are more directly determined by genetic makeup than by learned response.

THE GENUS

Basically, the genus is a group of related species that occupy broadly similar ecological niches. In spite of their close phylogenetic relationship, chimpanzees and man are not placed in the same genus because they occupy very different ecological niches. On the other hand, the widespread baboons are usually placed in a single genus separate from their close relatives the drills, because they generally live in savannahs, not in the forests like the drills.

The genus shares some of the importance of the species as a basic unit in evolution and taxonomy. Because nature has not created pigeonholes with neat boundaries for her life forms, species and genera tend to merge conceptually. For example, the reproductive isolation between the species of some genera is not complete. Hence, in comparative studies or in defining the limits of the behavioral repertoire, it is sometimes more meaningful to use the genus. Again, baboons are a significant example. Despite the lack of agreement on the number of species of baboons, it is generally agreed that the savannah-dwelling baboons are a single major adaptation and that all populations belonging to the genus *Papio* are closely related regardless of the number of species.

HIGHER CATEGORIES

The higher categories of taxonomy cannot be defined in any single manner, nor is it the purpose of this book to do so. For a thorough discussion of the problems involved in familial and ordinal classifications and their divisions, see Simpson (1961), Blackwelder (1967), and various other authors.

Very briefly, each group of animals (and plants, for that matter) must be examined to determine as closely as possible what the natural groupings are. Ideally, these natural groupings should form the taxonomic categories at all levels. The order, a higher category in the taxonomic system, is usually defined by prevailing evolutionary trends. For example, the order Perissodactyla (horses) is characterized by the evolution of running quadrupeds that tend to be adapted to grazing on grass. The early Perissodactyla were neither efficient runners nor grazers, but the history of the order since Paleocene times shows a series of changes that culminate in the modern horse.

Species and genera are classified into an order on the basis first of their shared common ancestry and second of their shared manifestation of the general trends of the order. We will be dealing only with the order

Primata. It must be borne in mind that other orders of animals must be classified rather differently since each order has its unique history of adaptation and evolution.

THE PRIMATES

Although the order Primata is a difficult one to characterize precisely, it is generally considered to consist primarily of arboreal animals that climb by grasping limbs and branches rather than by digging claws into the wood. But by no means do all primates live in trees (although the vast majority of species do), and some are still capable of climbing by means of their claws.

In the evolution of primates a series of major trends are noticeable that apply to many but not to all primates. Le Gros Clark (1959) discusses these trends in detail. Basically, they point toward (1) an orientation of sight rather than of smell as the major means of sampling the environment, (2) greater and greater reliance upon learning as a means of adapting to the environment, (3) a variety of locomotor systems but without the loss of a primitive flexibility in the limb joints and of the adaptation to climbing by grasping, and (4) the stabilization of the dentition with a grinding molar/premolar system.

MAJOR DEFINING CHARACTERISTICS OF PRIMATES

A wide variety of characteristics—ranging through anatomical, physiological, karyotypic, embryological, behavioral, and others—is used for solving taxonomic problems. A detailed and complete classification must consider and weigh all of the factors that contribute to an understanding of phylogenetic relationships. But classification can serve more limited purposes, and since our purpose here is somewhat limited, only the following major characteristics will be used: locomotion, special senses, diet, social organization, and habitat. Social organization and habitat are discussed in later chapters but locomotion, special senses, and diet will be introduced here.

Locomotion

As tree-dwellers, primates have evolved a number of interesting adaptations that enable them to exploit their arboreal habitat in various ways. The most generalized form of locomotion used by primates, a primitive quadrupedalism, does not differ markedly from the basic mammalian quadrupedalism of Mesozoic times before the great adaptive radiation of mammals in Cenozoic times. All four limbs retain the full complement of digits and a general flexibility of the limb joints (shoulder, elbow, wrist, hip, knee, ankle) that in other mammalian forms has often been modified by increasing restrictions. In fact, joint flexibility was actually enhanced during Cenozoic primate evolution, even among quadrupeds. Quadrupedal locomotion is effective principally for animals that move over surfaces and are supported by their limbs; it is less effective for animals that are suspended under surfaces. Few quadrupeds have become hangers; sloths are the major exception, and among the primates, lorises move under branches quadrupedally for short distances. One major characteristic of primate quadrupeds is their retention of a long tail for balancing the body during movement through the trees and for repositioning the body during leaps from branch to branch. Ground-dwelling quadrupedal primates have shorter tails.

Lorises, the slow-moving prosimians, represent a specialized form of quadrupedalism. These slow, careful animals never leap from branch to branch and virtually never have more than one or two of their four hands out of contact with the substrate. Since their problems of balance and body positioning do not require a tail, they have lost it during the evolutionary process.

Chimpanzees and gorillas have redeveloped quadrupedalism after having been adapted to arm-swinging under branches, a form of locomotion called brachiation. Since their arms and hands changed to adapt to brachiation, their return to quadrupedalism required new adjustments. They walk on their knuckles—supporting their weight on the metacarpals and first and second phalanges—instead of on their palms, and their shoulders are higher than their hips because of their long, brachiating arms.

Thus the three forms of quadrupedalism among primates are the generalized mammalian form tending to great joint flexibility, the loris form, which has even greater flexibility than the general form, and the form based on brachiation.

Saltation, or Vertical Leaping and Clinging Numerous primates have adapted to arboreal life by becoming saltators, leaping from one upright branch to another rather than moving along horizontal branches. The retention of the primate climbing by grasping in a saltating adaptation raised a problem kangaroos or other ground-dwelling saltators do not face. Kangaroos have developed the long foot for proper leverage in leaping by elongating the metatarsals in the middle of the foot. But the kangaroo adaptation separates the big toe from the other toes so far that it is useless as a grasping organ. The primate saltators have elongated the bones in the heel of the foot—the tarsals—and have thereby retained the relationship between the big toe and the other toes. Most leaping primates have also retained a long tail for repositioning the body in midair.

Semibrachiation The term "semibrachiator" is applied to primates that combine quadrupedal traits with brachiating traits. They move through the branches, using either the broad paths afforded by the major limbs or suspended beneath smaller branches somewhat like a full brachiator.

Their habit of dropping through the branches of the trees is facilitated by their ability to catch and swing under branches.

Brachiation   True brachiation as a major means of locomotion is the result of habitual movement under the limbs. The shoulder girdle, elbow, and wrist are extremely flexible, the arms and hands are elongated, and the arms are more powerful than the legs. Gibbons, the only living primate whose primary mode of locomotion is brachiation, move very rapidly under branches by arm-swinging or use their hind limbs to walk bipedally on top of the branches. Orangutans move slowly and carefully through the trees in a similar manner but use their legs as much as their arms to hang from branches and have evolved great flexibility in their legs also. Gorillas and chimpanzees, which have already been mentioned as practicing a form of quadrupedalism, hang from branches and chimpanzees brachiate; but today both are primarily quadrupedal.

Bipedalism   Man is the only habitually bipedal primate. His spine, pelvic girdle, legs, and feet are highly specialized to support the upright torso totally on the hind limbs and to stride over long distances. Most other primates can occasionally walk bipedally, but only the gibbon does it with apparent ease.

Special Senses

At some time in the Eocene epoch, primates began a special-sense revolution that brought some members of the order to a new adaptive plateau; there has been little apparent change in the special senses of the higher primates since that time. As a result, the more primitive primates are closer to the generalized mammalian tendency to emphasize smell over sight, whereas the higher primates have the form characteristic of the higher adaptive plateau.

Many mammals depend chiefly on their sense of smell. The snout is long and the sensory receptors for smell are large. The olfactory lobe in the brain is enlarged and forms, in some groups of mammals, a major part of the brain. Early primates and extant primitive primates also tend to have larger receptors and olfactory lobes than the more modern primates. But even among them there is a tendency toward a reduction in these areas.

Special Senses of the Higher Primates   The special-sense revolution that produced the higher primates was the result of a strong emphasis on the sense of sight. The visual apparatus—the eye as receptor and the portions of the cerebral cortex that receive visual impulses and store visual memories—became markedly enlarged with a concomitant reduction in the olfactory apparatus. This was probably the major impulse for the expanded neocortex in primates. Stereoscopic vision, with its more efficient depth perception, became the rule, and various degrees of color vision were developed in the diurnal higher primates.

Diet

Primates have a variety of dietary adaptations. Primitive mammalian diets, often at least partly insectivorous, ranged from carnivorous to herbivorous. Primates have exploited most of the mammalian range, but in them most forms are specialized to some extent.

Insectivorous   Insects provide a rich source of energy and require little specialization in mammalian dentition, which needs only to crush the insects to prepare them for digestion. The primitive shearing and crushing teeth of early mammals were well suited for this purpose, and primates that subsist primarily on insects retain a dental form close to the tribosphenic molars of Mesozoic mammals.

Carnivorous   Meat-eating provides large amounts of protein in small portions of food. An intestinal system adapted to an insect diet needs little modification to utilize meat effectively as a major source of nourishment. Since grinding is not important in preparing meat for digestion, the dentition can remain fairly primitive. Thus many of the insectivorous primates are also partially carnivorous.

Vegetarian   Primates living in an arboreal habitat and surrounded by a wealth of vegetable food have tended to specialize in vegetarian diets rather than in any of the other food sources. As a result, the forms of vegetarian diets differ.

One of the more specialized vegetarian diets is that of the *leaf-eaters*. Their adaptation involves a sacculated stomach, where mature leaves are fermented and the tissue is broken down to release the nutriments. These leaf-eaters, which rarely eat any other foods, do not require young shoots and survive better than many other primates where drought occurs. There are some indications that they can even supply their water needs by eating mature foliage.

Other primates are fruit-eaters, relying strongly on fruit as a food source to the exclusion of most other sources. These *frugivorous* primates are not widespread since theirs is a narrow specialization.

The most common vegetarian adaptation is the ability to eat various foods but with a concentration on vegetable foods. Old World *omnivorous–vegetarians* usually have cheek pouches where they store food for later mastication; this adaptation enables them to gather quickly and masticate at leisure in a safe place. Unlike the leaf-eaters, they cannot subsist on mature leaves; but they eat insects and occasionally meat and thus broaden their diet and exploit a somewhat different ecological niche.

SUBDIVISIONS OF THE ORDER PRIMATA

The primary differentiation between the two suborders of primates—
Prosimii and Anthropoidea—derives from the special-sense organization.
The trend toward the dominance of vision among the members of the
suborder Prosimii is complete among the members of the suborder Anthropoidea. This does not mean that the visual capabilities of the higher
primates do not differ. At least some of the New World monkeys (De-
Valois and Jacobs, 1968) do not discriminate colors in the green–red
range nearly so accurately as the Old World monkeys, apes, and men.
But all have stereoscopic vision and some form of color vision. They also
have a remarkably expanded cortex, particularly in the visual areas; their
mechanisms of smell, both in the nasal region and in the brain, are strik-
ingly reduced and relatively small. Whereas some prosimians approach
these conditions, all fall short of attaining it.

In many other areas, the prosimians and the higher primates show
similar adaptations. Both have a wide range of locomotor systems, of
dietary adaptations, and of social behavior.

PROSIMIAN DIVERSITY

Since our emphasis here is on behavior and since only living forms can
be studied through the broad spectrum of behavior, only the living forms
will be considered here. The various groups of prosimians have been
divided into three infraorders: Lemuriformes, Lorisiformes, and Tarsiiformes.

Lemuriformes

Today the Lemuriformes are the most diverse of the three, encompass-
ing the primitive tree shrews (Tupaioidea), the more advanced quadrupedal lemurs (Lemuridae), the specialized hopping indris lemurs
(Indridae), and the specialized gnawers (Daubentoniidae). Tree shrews
are on the borderline between the primitive basal insectivore stock and
the primates. Some authorities place them among the insectivores (Mc-
Kenna, 1963), others among the primates (Le Gros Clark, 1959). Despite
the arguments and the sound bases on which they rest, the important
point is that, being borderline, tree shrews have the characteristics of
both groups. They are small arboreal forms with claws on all digits and
some prehensility in the hands. Although they can to some extent climb
by grasping, they often use their claws. They are squirrel-like in body
form, but their brain is slightly more oriented toward sight than smell.
Even if they are not actually primates, they most nearly represent our
concept of the earliest primates.

The quadrupedal lemurs of Madagascar are very much more monkey-
like than the tree shrews. Their vision approaches the stereoscopic form
found in monkeys and apes, and their sense of smell is reduced to a

PLATE 1   The black galago, a variety of *Galago crassicaudatus*, showing the long foot of the saltator, the pleated ears of the nocturnal animal relying heavily on hearing, and the relatively long snout common for many prosimians. (*Courtesy of the Oregon Regional Primate Research Center; taken by Harry Wohlsein.*)

moderate extent. In general, they resemble long-snouted monkeys. Their hands have nails rather than claws and their feet retain one claw. Their way of life is that of an arboreal omnivorous–vegetarian.

The indris lemurs are similar to the others but have a specialized form of locomotion. They hop from branch to branch and spend most of the time with their trunks in a vertical position. Their habitat is similar to that of the lemurs with whom they are often found in close contact in Madagascar.

The aye-aye, the only living member of the Daubentoniidae, represents the last surviving rodent-like primate. Its specialization is its gnawing incisors, which are used to open nutshells, to extract parasite larvae, and to bite through bark in search of grubs. The nocturnal aye-ayes apparently live in pairs but occupy individual nests built of foliage and twigs for one to several days.

PLATE 2   Black lemur, *Lemur macaco*, one of the diurnal prosimians of Madagascar. (*Courtesy of the Oregon Regional Primate Research Center, taken by Harry Wohlsein.*)

Lorisiformes

The Lorisiformes are represented by two major groups that are differentiated at the subfamily level. Lorises are slow-moving carnivorous–omnivorous animals, excellently adapted to climbing by grasping. They are nocturnal and tailless since their slow locomotion through the trees at night does not require the tail as a counterbalance. The second digit is

PLATE 3   The slow-moving nocturnal potto, *Perodicticus potto*, a Lorisiforme. (*Courtesy of the Oregon Regional Primate Research Center; taken by Harry Wohlsein.*)

often reduced and thus a greater span between the first and the remaining digits is achieved for more effective and secure grasping.

Galagos are small hopping animals that have retained their tails for body positioning as they leap through the air. Like the lorises, they are nocturnal and have a similar diet. They probably rely heavily on insects in the wild. Their sense of hearing is specialized so they can accurately determine the direction of a sound and the distance of its source; this is a great advantage for catching insects.

Tarsiiformes

The single extant genus of Tarsiiformes, *Tarsius*, is also highly adapted to saltation as a means of locomotion. Tarsiers, however, have not specialized their hearing as much as the galagos but have strongly emphasized sight instead. Otherwise the two forms are rather similar.

ANTHROPOIDEA

The higher primates are generally subdivided into three superfamilies: the Ceboidea, or New World monkeys; the Cercopithecoidea, or Old World monkeys; and the Hominoidea, or apes and men. All share a special-sense system that includes stereoscopic color vision, a reduced sense of smell, hearing that is relatively sensitive but not specialized for depth perception, and a well-developed sense of touch and taste. Neither touch nor taste is specialized, but each has some degree of discrimination. At least in much of their resting time, higher primates tend to hold their torso with the vertebral column perpendicular to the earth's surface. Quadrupedal monkeys sit a great deal of the time and thus place their bodies in the vertical position. Although their locomotion generally places them in a horizontal position, their locomotion through the trees—climbing in and out of trees and up slanting branches—shifts the torso into a partially or fully upright position. The brachiating apes, which usually support themselves by hanging from branches, are even more likely to be found with their torsos in the vertical position. Sitting is a major resting position for both monkeys and apes. As a result, selection has shifted the mesenteries supporting the internal organs forward so that the latter are adequately supported in the vertical position. Sitting, common in the lower primates, is universal among the Anthropoidea. The suborder lacks only specialized saltation among the forms of locomotion, and diets tend to be highly vegetarian.

Ceboidea and Cercopithecoidea

Both the Ceboidea and Cercopithecoidea are classed as monkeys, the evidence indicating that they are a parallel evolution of the higher primate adaptation. Both have adapted thoroughly to the arboreal life typical of primates. The Old World forms have evolved a few ground-dwelling representatives: the baboons, patas monkeys, gorillas and humans of Africa, and some macaques in Asia; no such ground-dwelling forms have evolved in the New World monkeys. By Miocene and Pliocene times in Africa, extensive savannahs offered ample opportunity for the primates to exploit. Whatever open grasslands may have been present in South America were certainly far less extensive, and it may well be that the ecological niche had already been filled by nonprimates. In any event, all the New World Ceboidea have remained exclusively tree-dwelling, descending to the ground only rarely.

PLATE 4   Male bonnet macaque, *Macaca radiata*. Note the long tail for this species.

The Ceboidea include two different families and a few intermediate forms: the Cebidae, which are remarkably monkey-like in their adaptation, and the Callithricidae, which have several squirrel-like adaptations including the spiral climbing—circling around the trunk as they ascend —made possible by the claws on their hands and feet. The Callithricidae brains have advanced far beyond those of the prosimians, however, and they are therefore considered monkeys in a specialized form. The Cebidae are the only primates with prehensile tails, but not all Cebidae have them.

PLATE 5   Formosan macaque, *Macaca cyclopsis,* seated showing his grasping feet and the use of the tail as a prop and balancing organ.

The Cercopithecoidea are divided by dietary adaptations into two families: the Cercopithecidae, which have a simple stomach similar to our own and an omnivorous diet tending primarily to vegetarian; the Colobidae, which are leaf-eaters with a sacculated stomach. Locomotion is quadrupedal with some semibrachiation among the Colobidae.

Hominoidea

The Hominoidea are characterized by their specialization as brachiators. Even man is included in the group because, among other things, he has the shoulder structure of a brachiator rather than a quadruped.

The subdivision of this superfamily into families is based primarily on locomotor adaptation. The Pongidae, or great apes, have specialized strongly as brachiators, whereas the Hominidae, man and his immediate ancestors, have adopted bipedalism as their major form of locomotion.

PLATE 6   Crab-eating macaque, *Macaca fascicularis,* one of the smaller, long-tailed species. (*Courtesy of the Oregon Regional Primate Research Center; taken by Harry Wohlsein.*)

A LIMITED PRIMATE CLASSIFICATION

Tables 1 through 5 give a classification centering on those species that are mentioned in this book, that is, those about whose behavior we have some evidence either from field or laboratory studies. This classification does not, in any sense, pretend to be exhaustive or to solve the many problems of primate taxonomy; it is simply an attempt to provide an overview of the relationships between the species studied here.

The characteristics given in the tables are only a limited extract of the taxonomic characters used to establish the classification.

Figure 1 shows the general evolutionary history of the primates as it is often presented today. A preprimate basal stock in Cretaceous times probably gave rise to the earliest primates. Adapting to an arboreal life by climbing-by-grasping, the early primitive primates radiated into a wide variety of forms in Paleocene and Eocene times. Sometime during the Eocene epoch, rodents began to compete with primates, and many of the primitive primates became extinct during late Eocene times. The few survivors were further selected by competition with the diurnal arboreal

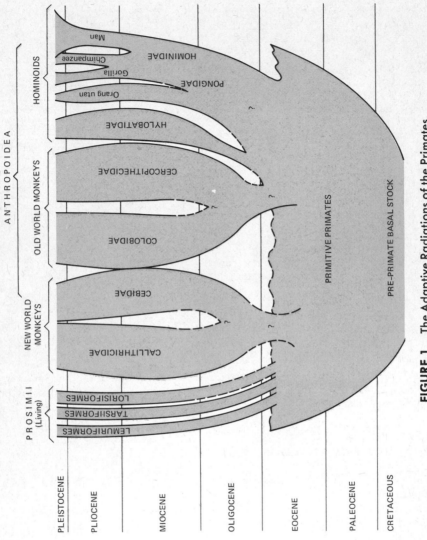

FIGURE 1 The Adaptive Radiations of the Primates.

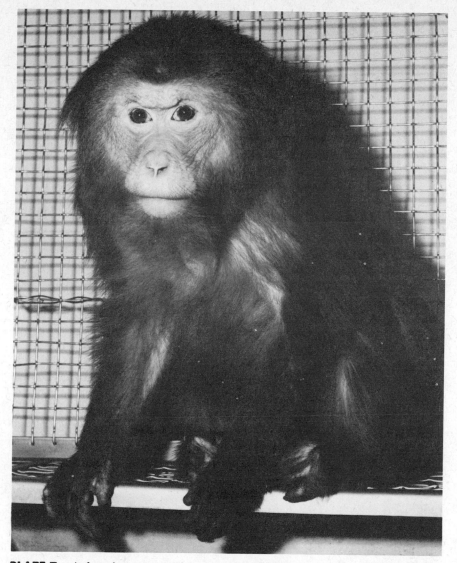

PLATE 7   A female stump-tailed macaque, *Macaca arctoides*. This is one of the large-bodied, short-tailed macaques from Southeast Asia. (*Courtesy of the Oregon Regional Primate Research Center; taken by Harry Wohlsein.*)

monkeys, which must have been evolving in late Eocene times; and as a result all living prosimians, with the exception of some of those living on Madagascar, where there are no monkeys, are nocturnal.

The primitive primates perfected the special senses of the higher primates during Eocene times, and once again there was a radiation of primates as they moved into the monkey ecological niche. This happened at least twice: once in the Old World to produce the Old World monkeys (and perhaps including the line leading to apes and men), and once in

TABLE 1. The Higher Taxa of the Living Primates

Order	Suborder	Infraorder	Superfamily	
PRIMATA — Climbing by grasping with nails instead of claws.				
	PROSIMII — Special senses close to basic mammalian senses but with different degrees of emphasis on vision.	LEMURIFORMES — Differentiated by the character of the auditory canal, brain, dentition, reproductive system, and other traits.	TUPAIOIDEA	Primitive insectivorous primate.
			DAUBENTONIOIDEA	Specialized in having large gnawing incisors.
			LEMUROIDEA	Procumbent incisors.
		LORISIFORMES — As above.	LORISOIDEA	Differentiated by the character of the auditory canal from lemurs and tarsiers.
		TARSIIFORMES — As above.	TARSIOIDEA	Advanced vision brain, nocturnal vision.
	ANTHROPOIDEA — Special senses with strong emphasis on vision (stereoscopic and color) with marked reduction of the sense of smell.		CEBOIDEA	New World quadrupedal and semibrachiating monkeys.
			CERCOPITHECOIDEA	Old World quadrupedal monkeys with bilophodont molars.
			HOMINOIDEA	Old World tailless brachiating apes and man with dryopithecine molars.

TABLE 2. The Studied Species of the Prosimii

Family	Subfamily	Genus and Species
TUPAIIDAE	TUPAIINAE	*Tupaia glis* Tree shrews: small *montana* shrub-dwelling *longipes* quadrupedal pri- *chinensis* mates with a diet *minor* of fruit and insects.
DAUBENTONIIDAE	DAUBENTONIINAE	*Daubentonia madagascariensis* Aye-aye: nocturnal, arboreal quad- ruped with a dietary preference for wood-boring grubs. Nests.
LEMURIDAE	LEMURINAE	*Lemur catta* Ring-tailed lemur *macaco* Black lemur Diurnal quadrupeds; the black lemur is arboreal but the ring-tailed comes to the ground; frugivorous and her- bivorous.
		Lepilemur mustelinus Sportive le- mur: nocturnal, arboreal, nesting sal- tator with a vegetarian diet.
	CHEIROGALEINAE	*Cheirogaleus major* Dwarf lemur: nocturnal, arboreal, nesting quad- ruped. Has periods of torpor ap- proaching hibernation. Fruit and in- sect eater.
	INDRISINAE	*Avahi laniger* Avahi: nocturnal, ar- boreal saltators probably having a diet solely of vegetation.
		Propithecus verreauxi Sifaka: diur- nal, arboreal saltators. Apparently en- tirely vegetarian.
LORISIDAE	LORISINAE	*Loris tardigradus* Slender loris: slow, arboreal quadruped, nocturnal with an insectivorous and carnivorous diet.
	GALAGINAE	*Galago senegalensis* Bushbabies *crassicaudatus* Thick-tailed galago Nocturnal, arboreal, nesting saltators with a diet of insects and vegetation.
TARSIIDAE	TARSIINAE	*Tarsius syrichta* Philippine tarsier: Nocturnal, bush-dwelling and arbo- real saltator, insectivorous and car- nivorous.

TABLE 3.   The Studied Species of the Ceboidea

Family	Subfamily	Genus and Species
CALLITHRICIDAE	....	*Saguinas midas* Red-handed tamarin: clawed, arboreal quadruped. Diurnal, nesting frugivore.
CEBIDAE	AOTINAE	*Callicebus moloch* Dusky titi: quadrupedal but also a saltator. Diurnal and arboreal. Frugivorous, insectivorous; also eats birds and eggs. *Aotus trivirgatus* Douroucouli or night monkey: nocturnal, arboreal quadruped. Diet of fruit, insects, small animals.
	ALLOUATTINAE	*Allouatta villosa* Mantled howler monkey: arboreal, diurnal, prehensile-tailed quadruped. Also considered a semibrachiator. Vegetarian. Greatly enlarged hyoid bone for howling characteristic.
	ATELINAE	*Lagothrix cana* Woolly monkey: arboreal, diurnal, prehensile-tailed quadruped. Also considered a semibrachiator. Vegetarian. *Ateles geoffroyi* Spider monkey: arboreal, diurnal, prehensile-tailed quadruped. Also considered a semibrachiator. Mainly frugivorous.
	CEBINAE	*Saimiri sciureus* Squirrel monkey: small arboreal quadruped. Diurnal. Diet of fruit and insects and probably some meat.

TABLE 4.   The Studied Species of the Cercopithecoidea

Family	Subfamily	Genus and Species

Erythrocebus patas   Patas monkey: terrestrial running quadruped.

Cercopithecus aethiops   Vervet
       ascanius   Red-tailed monkey
Vervet semiterrestrial, red-tailed arboreal; both quadrupedal.

Cercocebus torquatus   White-collared
                                      mangabey (semiarboreal)
         albigena   Grey-cheeked
                                      mangabey (arboreal)
Quadrupedal monkeys.

CERCOPI-
THECINAE

Quadrupedal omnivorous vegetarians.

Theropithecus gelada   Gelada: terrestrial sitting quadruped, vegetarian diet supplemented by insects.

Papio hamadryas   Hamadryas baboon
       anubis
       cynocephatus   Savannah baboon
       papio
       ursinus
Terrestrial walking quadruped.

CERCOPI-
THECIDAE

All diurnal.

Cynopithecus niger   Black ape of Celebes
Perhaps semiarboreal, quadruped.

Macaca fascicularis   Crab-eating macaque
       fuscata   Japanese macaque
       mulatta   Rhesus macaque
       nemestrina   Pig-tailed macaque
       radiata   Bonnet macaque
       silenus   Lion-tailed macaque
       sinica   Toque macaque
       sylvana   Barbary ape
Semiarboreal quadrupeds with variable terrestrial adaptation.

COLOBINAE

Quadrupedal with sacculated stomachs.

Presbytis entellus   Hanuman langur
       johnii   Nilagiri langur
Semiarboreal and arboreal quadrupeds.

Colobus polykomos   Guereza: arboreal quadrupeds.

TABLE 5.   The Studied Species of the Hominoidea

Family	Genus and Species
HYLOBATIDAE Diurnal.	*Hylobates lar* Gibbon: diurnal, highly arboreal, swift brachiator. Omnivorous–vegetarian. Little sexual dimorphism in size.
PONGIDAE Diurnal, use sleeping nests.	*Pongo pygmaeus* Orangutan: diurnal, cautious arboreal brachiator. Predominantly frugivorous. *Pan satyrus* Chimpanzee: diurnal, terrestrial quadruped and arboreal brachiator. Omnivorous–vegetarian. *Gorilla gorilla beringei* Mountain gorilla: diurnal, terrestrial quadruped modified from a brachiator. Vegetarian.
HOMINIDAE	*Homo sapiens* Man: terrestrial biped modified from a protobrachiator. Omnivorous–carnivore.

South America to produce the New World monkeys. The advances were so well adapted to the primate arboreal way of life that selection tended in that direction in more than one line of early primates. Further specialization differentiated the leaf-eating langurs and colobus monkeys from the more omnivorous Cercopithecinae and eventually man from his ape-like ancestor.

Chapter 2
Primate Evolution and Distribution

The ecology of an organism, which deals with the mutual relationships between an organism and its surrounding environment, is an important factor in the social organization of a species. The effect on the environment of societies of animals is different from that of single animals of the same species, and these societies are, in turn, molded and altered by the changing ecology. The ecology is the setting within which the society operates, and it impinges, sometimes directly, sometimes subtly, on the workings of the society.

Primate societies are one of the major means by which primate species have adapted to their environment. Virtually no species of primates exists without a social organization of some sort, even if it is transitory and restricted to mating and rearing the young. Most primate societies are composed of numerous individuals and characterized by a complex organization. How these societies function is the major topic of this book. But the setting in which they function and to which they adapt must also be examined if one is to understand the reasons for many of the interesting aspects of the society. Hence, a complete chapter (Chapter 11) will be devoted to the interacting relationship between the environment and the society.

The ecology of a species involves a number of factors. One of the first to consider is its geographic distribution together with the climate and the flora of the regions inhabited by the species. Narrowing the discussion to the local environment, one is first concerned with the specific habitat occupied by the species, whether trees, plains, open woodland, or other particular niches. Diet is another important consideration, since the sources of food determine which parts of the habitat are the most important, but havens from predation are also a necessity.

No species of animals occupies a geographic area, a particular kind of forest, and eats specific foods in isolation from other animals. This essentially social aspect of living in the wild sets up a series of dynamic relationships that determine to a large extent the distribution of species. There are, first of all, predator and prey relationships that limit the distribution and behavior of the species, whether of the predators (for example, lions) or of the prey (for example, many primates). Added to this is the competition for food between forms that are not in a predator–prey relationship, as well as the cooperative relationships between species. Finally, there are the omnipresent parasites that have adapted to living on their hosts without killing them.

DISTRIBUTION OF PRIMATES

A brief survey of the past ecological and geographical distribution of primates is a good basis for a discussion of present primate distribution and ecology. Such a survey shows the general adaptations made by primates during their evolutionary history and indicates some of the changes necessitated by these adaptations and the capacity for survival of the more specialized or divergent of the modern forms, among them the baboons and man.

With the exception of a handful of enigmatic Cretaceous teeth, the earliest known primates came from the Paleocene fossil beds, most of them during the later part of the Paleocene epoch, that is, between 65 million and 56 million years ago according to present estimates. These earliest fossils have so far been found only in North America and Europe; several reasons for this restricted distribution have been suggested. First, only a few small mammal fossil beds from Paleocene times have been worked in other parts of the world; when they are worked, they may yield evidence of primate presence. Second, the increasing evidence of continental drift indicates that the Atlantic is a relatively young ocean. Some 100 million years ago, that part of North America east of the Rocky Mountains and of Europe west of the Urals may well have been a single continent; the similarity of primate and other faunal elements in these regions suggests that they were a single zoogeographic area. The Tethys Ocean, of which the Mediterranean Sea is a remnant, and recurrent shallow seas blocked the expansion southwards into Africa and South America, so that the homeland of the primates may well have been an ancient Euramerican continent or the larger Laurasian continent. Their spread into other places must have occurred after separating portions of the old continent made contact with new land masses or the shallow seas receded. The increasing knowledge of such geologic processes and of the history of the oceans and land masses should help immeasurably to explain the distribution of primates (Bullard, 1969; Tarling, 1971).

Paleocene times were warmer than our present epoch in the northern continents, which were then in the tropical and subtropical latitudes. South America and Africa were closer to the south pole but the ocean currents were less restricted than they are today and the climate over the whole north was moderated by them. Thus millions of years ago Europe and North America had tropical climates, and the primates that lived there had no temperate seasonal climates to adapt to. If primates were spread throughout much of the earth's land mass (an hypothesis that is less and less likely in light of the new evidence of land movements cited above), the Paleocene climate was apparently able to support more of the tropical forests to which most primates have been and are adapted than the present climate does.

The older interpretations of the fossil record have it that the northern extension of primates into Europe and North America during Paleocene and Eocene times probably represents the extreme limits of primate distribution rather than the center of it. According to this older interpretation, it is highly unlikely that the European and North American fossils studied today are the remains of the ancestors of any living forms of primates. What we know of Paleocene primates is probably limited to the aberrant and the rare, to the peripheral species rather than to the central core of the Paleocene primate population. In this latter interpretation, climatic change was the stimulus of evolutionary change, which was followed by readjustments of the adapting fauna and flora.

If, on the other hand, continental relationships changed with the continental drift mentioned above, the faunal and floral assemblages that had evolved on separate continents (for example, the hypothetical Euramerica) would have been faced with the intense pressure of contact with other such assemblages when the European segment touched Africa and Asia and the North American segment contacted South America. Under these circumstances, the primitive primates would have been presented with new and changing ecologies to which they would have been forced either to adapt or to succumb. This kind of stimulus may explain, at least partially, the evolutionary changes that took place in the first half of the Cenozoic era. Moreover, our interpretation of primates as the rodents of the Paleocene epoch may be misleading; they may represent the rodent niche only in ancient Euramerica, and true rodents may have evolved on some other ancient land mass.

MAP 1   Land Masses around the central Atlantic about 100 million years ago. Shallow seas cut off Africa and South America from the northern continents and were beginning to separate North America and Europe. (*Adapted from D. H. and M. P. Tarling,* Continental Drift, A Study of the Earth's Moving Surface, *London: G. Bell and Sons, 1971; New York: Doubleday. Base map courtesy of Goode's Series of Base Maps, Department of Geography, The University of Chicago. Copyright by the University of Chicago, Department of Geography.*)

PALEOCENE PRIMATES

Paleocene primates were primitive mammals; none could be called monkeys or apes in the modern sense of the terms. Most of them would seem backward when compared with even living lemurs, lorises, and tarsiers. Their generalized mammalian characteristics placed them close to insectivores and other primitive mammals. Their teeth were of the primitive tribosphenic (rubbing wedge) variety with few of the specializations seen in living primates. They relied strongly on the sense of smell; the sense of sight had not assumed the importance it has with most extant primates. Nevertheless, several features indicated their emergence into the primate condition. An occasional flattened terminal phalange supporting a nail indicates that although Paleocene primates were primarily clawed, they had already begun to climb by grasping. Despite a still highly generalized dentition differing only little from that of most other primitive mammals, some features that could easily be ancestral to modern primate dentition had begun to appear. In other words, these extremely primitive and generalized mammals were about to produce the morphological characteristics found in the higher primates today.

Three groups of Paleocene primates from ancient Euramerica represented by fossil remains today are the families Carpolestidae, Plesiadapidae, and Phenacolemuridae, none of which can be placed in an ancestral relationship to any of the living primates. Their general character was that of primitive lemur-like animals that climbed by grasping and by using their claws. Small animals, many no bigger than a large rat, they were primarily, perhaps entirely, tree dwellers. Omnivorous to a certain extent since their dentition still had many generalized features, they probably relied heavily on insects and fruit. Some may have already begun to specialize in their diet; for example, the loss of canines among the Phenacolemuridae indicates a somewhat specialized dentition.

EOCENE PRIMATES

The fossil record shows that by the Eocene epoch the early primates had attained their greatest florescence and the widest range of ecological niches. The lines of evolution leading to the modern primates may well have been laid down by the end of Eocene times. Most of the fossil material of Eocene primates comes from North America and Europe; the only exceptions are two intriguing late Eocene fragments of jaws from Burma, *Amphipithecus* and *Pondaungia*, considered from their dentition to be possible pongids. Even without fossil records from Africa, South America, and Asia, Eocene primates fall into six families rather than three: Tarsiidae, Adapidae, Anaptomorphidae, Microsyopidae, Omomyidae, and, perhaps, Pongidae.

Some features of some Eocene groups could relate them to living populations. The Omomyidae, for example, a European family of primates, had a dentition somewhat resembling the squared, grinding molars of modern monkeys and apes. Others are considered ancestral, or at least closely related, to the living tarsiers because of their enlarged eye orbits, which are nearly fully enclosed posteriorly. The Tarsiidae were a major and apparently widespread group during Eocene times. Some of the other families may well have been ancestral to living lemurs although the connection between them is still tenuous. Again, the problem is that we are dealing with North American and European forms that may have been either peripheral to the main lines of primate evolution or represented those lines before they were forced into the evolutionary crucible of interregional contact. When African, South American, and Asian Eocene deposits are better known, we may discover forms that are closer to the ancestral populations of living primates; on the other hand, increasing knowledge of the periods of contact between zoogeographic areas may reveal the connections between the earlier and more modern forms.

According to the present evidence, none of the Eocene primates were monkeys or apes. But what about those tantalizing jaw fragments from *Amphipithecus* and *Pondaungia*? Their lower molar teeth with low cusps resemble those of living apes and men; this dental evidence could identify them as the earliest higher primates. We cannot, of course, deduce anything about their special senses from their teeth or assume that either genera had other features recognizable as anthropoid. Although some of the adaptations that produced the higher primates probably began in Eocene times, many changes in locomotion and the special senses occurred later.

During Eocene times, the prosimians were at their zenith: they had begun to spread over the continents of the Old World and were highly diversified. Although most of them were probably arboreal, some may have lived on the ground or at least in low bushes and there is even some indication that a few burrowed into the ground. Many developed the large gnawing incisors typical of rodents; however, the primate incisors, unlike the rodents', were not ever-growing and could not be replaced when worn too far. These prosimians probably occupied many of the niches exploited today by rodents, which had not appeared in full force in Europe and North America by early Eocene times. Some were probably evolving in the monkey direction, growing bigger brains rather than bigger teeth and becoming more at home in the trees. In the process they were probably changing from small scurrying animals like the living insectivores to animals with a slower and more relaxed way of life. Their metabolisms likewise were probably slowing down from the frenetically active pace characteristic of the insectivore ancestors of primates to the more leisurely one of modern mammals. As a result, they probably spent less time searching for food and more time indulging in the complexities of social behavior, a considerable advantage for many reasons: cooperative protection of the group, more flexibility with which to meet different conditions through changes in social organization, and so on. However, such prosimians probably constituted only a small group. The mainstream of primate evolution in the Eocene epoch consisted of the small, numerous, and varied prosimians; the monkey-like tendencies of the few might well have been considered by an observer to have been aberrant, perhaps overspecialized, and not likely to be of much importance in the long run. Speculation aside, we still have no fossils that could by any stretch of the imagination be identified with monkeys.

OLIGOCENE PRIMATES

After the rich diversity of the Eocene primate fossils, those of the Oligocene are disappointing. Only two localities have produced the fossil remains of Oligocene primates, the Fayum basin in Egypt and two sites in Mongolia. A handful of mandibles and a couple of fragmentary skulls, all but one mandible coming from the Fayum basin, constitute the scanty evidence for over 10 million years of evolution.

However, even these few fragments reveal something about the progress of primate development. Some of the specimens have monkey-like dentition; others, ape-like dentition. *Parapithecus* and *Propliopithecus* of the Fayum have grinding molars that are closer to those of the modern gibbons than to any other living forms. *Aegyptopithecus* and *Oligopithecus* of the Fayum and "*Kansupithecus*" from Mongolia have dentition rather similar to the Eocene *Amphipithecus* of Burma, which some authorities (Piveteau, 1957) suggest are thus close to the ancestral line of living great apes and man. Another form, *Moeripithecus,* has molars tending to the bilophodont or two-ridged condition of modern monkeys.

The scanty material from the Oligocene deposits points to the development of four major branches of higher primates—gibbons, Old World monkeys, New World monkeys, and the great apes—which may already have been separated and were developing in their own directions. The presence of *Amphipithecus* in Eocene Burma, *Oligopithecus* and *Aegyptopithecus* in Oligocene Africa, and "*Kansupithecus*" in Oligocene east Asia suggests that the apes were already widespread by Oligocene times. These ancestral apes were certainly small compared with the modern chimpanzee, to say nothing of the orangutan and gorilla. No limb bones have yet been found for Oligocene primates, but the later Miocene and Pliocene remains indicate that the Oligocene "apes" must have been more like monkeys than like the modern apes in their locomotor adaptation. However, they probably had already begun to move toward a partially brachiating way of life somewhat resembling that of the modern Old World semibrachiators, which move their forelimbs through a greater arc than other Old World monkeys do.

In Oligocene times, the earth's climate had begun to resemble conditions as they are today. By late Pliocene times, the tropics had undergone the reduction to their present condition. This appears to have been a slow process, and primates were still living in parts of Europe during the Late Tertiary. But the extreme limits that primates had reached in Paleocene and Eocene times (if indeed these are limits and not simply a different positioning of the land masses) seem to have been sharply cut back by mid-Oligocene times. The few prosimians in North America at the opening of the Oligocene epoch were brief holdovers from the warmer, pre-mountain-building Eocene times. We have no evidence that any important primate evolution took place in Oligocene North America or Europe.

Rodents seem to have come into their first moment of glory during the Oligocene epoch and to have preempted many of the ecological niches that had previously supported the primitive primates. The ever-gnawing, highly prolific rodents evolved rapidly and spread widely, always relying on their ever-growing incisors to outgnaw their competition. As a result, food supplies must have been drastically reduced for the primitive primates, which could eat only the softer parts devoured by the rodents as appetizers for the more robust roughage. According to the statistics of modern rodent and primate reproduction, the rodents also outreproduced the primates, each female producing many litters per year instead of the one or two offspring that is usual for prosimians today. The upshot is that the primate remains of the Oligocene age, scanty though they are, consist mainly of forms that were already specializing in a kind of life similar to that of modern monkeys and apes rather than of forms that had to compete, directly and unsuccessfully, with rodents for food and living space. An observer in early Oligocene times would have been justified in consigning the primates to oblivion along with the other primitive orders of mammals that were replaced in later Teritary times by more advanced and intelligent mammals. A discerning eye was needed to see that some strange primates were moving off in new and promising directions.

MIOCENE PRIMATES

By the Miocene epoch, the apes had probably become widespread and relatively varied. As many as fifty genera of Miocene and Pliocene apes, exclusive of gibbons and *Oreopithecus,* have been proposed on the basis of fossil remains, but Simons and Pilbeam (1965) recently reduced this somewhat fanciful number to a more realistic two genera. There are numerous remains from the Miocene of Africa, Asia, and Europe. Certain signs in the elbow joints of the gibbonoid forms, *Pliopithecus* and *Limnopithecus,* indicate that selection had favored those that could brachiate. No shoulder remains, other than those of *Oreopithecus,* are extant from Miocene ape-like primates; this is unfortunate since they are better indicators of the extent of brachiation than elbows. Several species of *Proconsul,* or alternatively *Dryopithecus,* from Africa could well be ancestral to the living chimpanzee and gorilla. The European and Asian species of *Dryopithecus* show that this genus was widespread in Miocene times.

The dentition of the Italian *Oreopithecus* differs somewhat from that of other apes, but its limb bones indicate that it was by no means a quadrupedal monkey. It has been said that *Oreopithecus* would make a good ancestral biped. But other claimants for the position of early hominid have stronger claims than *Oreopithecus* because of their dentition. The dental remains of African and Indian *Ramapithecus*, showing small canines and a parabolic dental arch, constitute the strongest claims from the Miocene–Pliocene epoch to being early hominid. But the evidence is too scanty to be conclusive.

The only Old World remains of the Miocene epoch that could have belonged to monkeys are a handful of teeth similar to those of the Pliocene *Mesopithecus*. Monkeys as we know them today must have evolved by Miocene times in the Old World; future excavation will probably unearth new fossils. According to dental evidence, apes, many of which were small creatures, were probably far more numerous than monkeys in the Miocene epoch; the monkeys probably represent a later radiation, although the stock had already diverged in the Oligocene epoch. The general assumption that monkeys are more primitive than apes and evolved earlier is based on the fact that apes have specialized locomotion and more elaborate brains, whereas monkeys have retained the more primitive and generalized quadrupedalism. However, brachiation in the apes was a later adaptation, although they may have tended in that direction early; fully adapted brachiators such as the gibbon and orangutan probably did not appear until late Pliocene times. Certainly the Miocene fossils do not show the same degree of specialization in brachiation as living forms. It is just as conceivable that populations ancestral or closely related to modern apes formed the major primate populations of Miocene times and were later superseded in many of their niches by modern monkeys as it is possible that the ancestral monkeys were widespread during the Miocene and were the basal higher primate stock.

Brachiation is sometimes regarded as an adaptation to feeding at the ends of branches. It should be remembered, however, that a small macaque clinging to one small branch can pick fruit as easily as a large orangutan clinging to three or four tips of major branches. Is it possible that the dentition of monkeys is better adapted to their dietary needs than that of the apes? Both New and Old World monkeys have bilophodont molars. This strong similarity may well have been advantageous to monkeys in competition with small apes, enabling monkeys to restrict the apes to the specialized niches they now occupy. A more complete fossil record may determine the answers.

Three genera of New World monkeys have been recognized in the Miocene deposits of South America: *Homunculus, Dolichocebus,* and *Cebupithecia.* Skulls and limb bones indicate that they were well on their way to becoming the modern New World primates. The skulls are not so large and capacious as the modern forms, but the slight differences can only indicate that they must have already been evolving in that direction for a considerable time.

No lemur fossils nor North American primate fossils of any sort have been found in Miocene deposits. It was not until terminal Pleistocene times that North America again played an important role as a primate habitat. By Miocene times, lemurs had probably been relegated to their present status as poor relations of the more advanced monkeys and apes. Although they may still have been relatively common in the reduced Miocene tropics, compared with monkeys and apes, they were probably already insignificant.

PLIOCENE PRIMATES

The Pliocene primate record of the New World is rather barren. A few early Pliocene specimens are similar to the Miocene forms, but most of the Pliocene fossils of the New World lack primate representation. However, primates in South America must have been completing the evolutionary process toward their present condition and diversity.

Most of the Old World Pliocene fossils, like those of the Miocene, represent apes. The gibbons are represented by the Asian form *Hylopithecus* and the anthropoid apes by a number of jaw fragments primarily from the Siwalik Hills of India. *Ramapithecus* and Leakey's recent finds of *Kenyapithecus* in Africa have helped to close the gap between man and some of his possible ancestors (Leakey, 1967).

Basically, what we know about the Pliocene apes is that they ranged from Africa as far as Mongolia. Thus they were more widely distributed than they are today. Apes were found in areas of the Sudan that are now habitable only by man and in Mongolia, where the climate now is too cold and dry to support the kind of forest necessary for modern gibbons.

The Pliocene Old World monkeys are better represented in the fossil record than the Miocene forms. An Egyptian form called *Libypithecus* is represented by a colobid skull. Apparently the modern leaf-eating monkeys of the Old World had evolved by this time, since nothing about this skull could be considered primitive. It does not represent any species living today, but it could so easily fit into the genus *Colobus* that it seems excessive to set up a separate genus for it. Another fossil from Europe, *Mesopithecus,* which also belongs to the leaf-eating colobids, indicates that the European climate was still warm enough to sustain the non-human primate population. Pliocene fossils of macaques and baboons demonstrate that the division of Old World monkeys into the present two families had already taken place. Although we have no fossil remains of sacculated stomachs, we can reasonably infer from *Libypithecus* and *Mesopithecus* that leaf-eating colobids, and from baboon and macaque fossils that the more omnivorous (though still primarily vegetarian) cercopithecids, were fairly widely distributed over the Old World tropics. The presence of *Mesopithecus* in Europe indicates again that the tropics of Pliocene times were more extensive than they are today. When *Mesopithecus* lived, the glaciations had not yet altered the climatic zones of the earth.

Lorisoids, nocturnal primate carnivores found today in Africa and south Asia, first appeared during the Pliocene epoch. The first loris fossil of essentially modern form was found in India. There are no indications that these interesting little animals existed during the Paleocene and Eocene epochs, the heyday of the prosimians. They may well be a late specialization of the prosimians, adapted to a niche that competes with neither the rodents nor the monkeys.

PRESENT PRIMATE DISTRIBUTION

American nonhuman primates are distributed today almost exclusively in the rain forests of the New World, from southern Mexico through Central America to northern Argentina in South America. Marmosets are practically confined to the Amazon basin, but the Ceboidea, including the best known, scientifically, of the New World monkeys, the howlers, extend beyond the barrier of the Andes in northern South America into the Central American region (Sanderson, 1957). Several other species are found in Central America.

The Old World monkeys and apes are found in Africa and southern Asia, with a gap from Arabia to India. Old World primates have by no means been confined to the tropical forests; man's adaptation to ground-dwelling is just one of many examples of primate migration from the forests onto open ground. Marginal forests, open savannahs, and even desert regions have been exploited by Old World primates. Living prosimians are found in the tropics of Africa and south Asia, with the highest concentration of species living on Madagascar.

The Apes

The gibbon is widespread throughout southeastern Asia from Assam to Hainan in the north through Malaya, Sumatra, Java, Borneo, and the related islands (Map 2). It is found only in the primary forests of these areas and is almost exclusively tree-dwelling. The other Asian ape, the orangutan, was found in vast areas of eastern Asia during Pleistocene times, but today it is restricted to the relatively undisturbed forest regions of Borneo and Sumatra (Map 2).

The chimpanzee is found in and around the tropical rain forests throughout Africa, from Sierra Leone through to the Congo basin. The pygmy chimpanzee (Pan paniscus) is separated from the others in the tropical forest of the bight formed by the Congo and Lualaba rivers. The gorilla population is divided between two widely separated areas, both in the Congo rain forest. The present distribution of the lowland gorilla in the west lies in the rain forest between the Ubangi and the Cross rivers. The mountain gorilla is confined to the region east of the Lualaba River and west of a line joining Lakes Albert and Tanganyika between the latitudes of 0° to 4°20′ South. According to Schaller (1963), the dis-

Hylobates species (gibbons)

Pongo pygmaeus (orang utans)

MAP 2   Present-day distribution of the Southeast Asian Pongidae. (*Base map courtesy of Goode's Series of Base Maps, Department of Geography, The University of Chicago. Copyright by The University of Chicago, Department of Geography.*)

tribution within this area is discontinuous. It is true of them as it is of most primates: They select an optimum habitat, leaving uninhabited areas lying within much of their range. There are some indications that in fairly recent times the distribution of the two groups was continuous before the forest shrank south of the Ubangi and Uele rivers in north central Africa (Schaller, 1963; Coolidge, 1936; Allen, 1939). (See Map 3).

■ *Pan troglodytes* (chimpanzee)

▤ *Pan paniscus* (pygmy chimpanzee)

▧ *Gorilla gorilla* (gorilla)

MAP 3   Present-day distribution of the African Pongidae. (*Base map courtesy of A. J. Nystrom and Company.*)

The Leaf-Eating Monkeys

The leaf-eating monkeys of the Old World, the Colobidae, form two discontinuous distributions. The African leaf-eaters, the *Colobus* monkeys, are confined to the forests of Africa (Map 4), both montane and tropical. No species of *Colobus* has adapted to low-bush or savannah dwelling. In southern Asia, from India through Southeast Asia into the Indonesian archipelago, different varieties of leaf-eating langurs are found that fall into three genera or species groups, depending upon the classifier: *Presbytis, Nasalis,* and *Pygathrix.* Most of them are confined to the forested regions—rain, mountain, or deciduous. At least one of these species, *Presbytis entellus,* has adapted to partial ground-dwelling; it is found in many parts of India in scrub forest, open woodlands, and in fields and

the Genus Colobus

MAP 4   Present-day distribution of the African Colobidae. (*Base map courtesy of A. J. Nystrom and Company.*)

towns. The subjects of recent studies, *Presbytis johni* in southern India and *Presbytis senex* in Ceylon are among the more arboreal langurs. Probably because it is partially ground-dwelling, *P. entellus* is far more widespread than any other species of langur; the other *Presbytis* species in southeastern Asia and the remaining genera of langurs (*Nasalis, Rhinopithecus,* and *Pygathrix*) are more restricted in their individual distributions (Maps 5, 6).

Omnivorous–Frugivorous Monkeys
The more omnivorous Old World monkeys, the Cercopithecidae, are found in the forests, the scrub, and the savannahs of the tropics and even of the temperate regions. They include the genera *Cercopithecus, Cer-*

Presbytis entellus
Presbytis johni (India)
Presbytis senex (Ceylon)
Pygathrix nemaeus

MAP 5   Present-day distribution of Asian Colobidae: I. (*Base map courtesy of Cartocraft Desk Outline Map, copyright © by Denoyer-Geppert Company, Chicago, used by permission; and Goode's Series of Base Maps, Department of Geography, The University of Chicago. Copyright by The University of Chicago, Department of Geography.*)

cocebus, *Erythrocebus, Theropithecus,* and *Papio* in Africa and *Macaca* in Asia. The present distribution and variety indicate that Africa was probably the continent of origin for these monkeys, which then contributed the macaques to Asia. In contrast, the leaf-eaters are varied and numerous in Asia, the genus *Colobus* evolving perhaps after a migration of leaf-eaters from Asia to Africa.

Remaining species of *Presbytis*

Nasalis

Rhinopithecus

MAP 6   Present-day distribution of Asian Colobidae: II. (*Base map courtesy of Cartocraft Desk Outline Map, copyright © by Denoyer-Geppert Company, used by permission.*)

Macaques   Macaques and baboons are of considerable interest in almost any study of primate social behavior because their ground-dwelling tendencies make them easily accessible. Except for the Barbary apes (*Macaca sylvana*), macaques are Asian and baboons are African (Maps 7, 8, 9). *M. sylvana* is isolated from the main distribution of living macaques and is found in the coniferous forests and hills of Algeria, Morocco, and, in the recent past, of southern Spain. The Spanish part

‖‖	*Macaca arctoides*	▓	*Macaca mulatta*
▨	*Macaca assamensis*	☰	*Macaca silenus*
‖‖	*Macaca radiata*	◪	*Macaca sinica*

MAP 7   Present-day distribution of macaques: I. (*Base map courtesy of Carto-craft Desk Outline Map, copyright © by Denoyer-Geppert Company, Chicago, used by permission.*)

of their distribution is now limited to the island of Gibraltar. Living macaques are found again only when one reaches India, although fossil macaques from France, Germany, and Italy (Piveteau, 1957) indicate they were distributed over a wider range and probably were at one time spread over the Near East.

The best known and most widespread of the macaques is *Macaca mulatta*, the rhesus monkey or bandar, which is common in the labora-

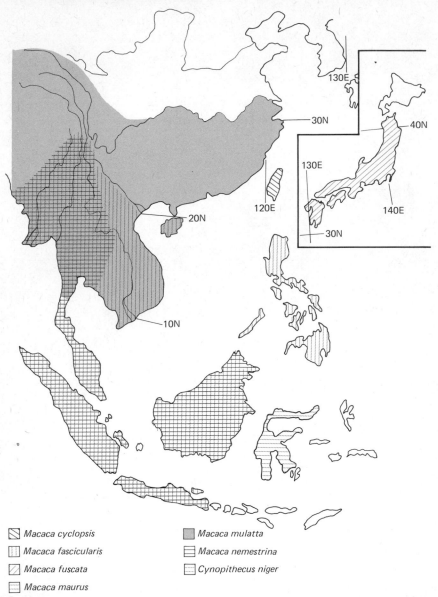

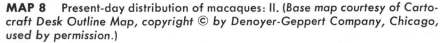

◨ Macaca cyclopsis	▨ Macaca mulatta
▥ Macaca fascicularis	▤ Macaca nemestrina
▨ Macaca fuscata	▦ Cynopithecus niger
▦ Macaca maurus	

MAP 8   Present-day distribution of macaques: II. (*Base map courtesy of Carto-craft Desk Outline Map, copyright © by Denoyer-Geppert Company, Chicago, used by permission.*)

tories of the United States and Europe. It is found in northern India, from the Godavary River north into Kashmir, extending eastward into Nepal, Bhutan, Sikkim, Burma, and down the Malay Peninsula to the latitude of Tenasserim (12° north). They have been reported in China and in parts of Tibet as far north as 35° north latitude. Their range in altitude is from sea level to 10,000 feet in Kashmir (Hinton and Wroughton, 1921).

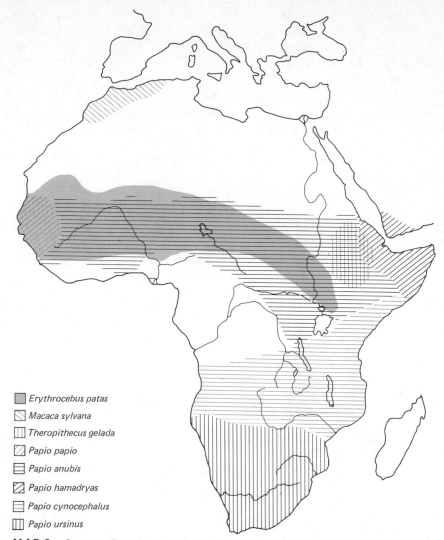

Erythrocebus patas

Macaca sylvana

Theropithecus gelada

Papio papio

Papio anubis

Papio hamadryas

Papio cynocephalus

Papio ursinus

MAP 9   Present-day distribution of ground-dwelling African monkeys. *Papio hamadryas* is referred to as the hamadryas baboon in the text, while the remaining species of *Papio* are referred to as the savannah baboons. (*Base map courtesy of Goode's Series of Base Maps, Department of Geography, The University of Chicago. Copyright by The University of Chicago, Department of Geography.*)

Entirely within the range of the rhesus macaques is the distribution of the closely related species *M. assamensis.* There is at present no indication of any adaptive difference between these species, but *M. mulatta* has only recently been studied in truly natural habits s and *M. assamensis* has yet to be studied. Only when a thorough study of *M. assamensis* is undertaken will there be a clearer understanding of their relationships. However, it is highly probable that *M. assamensis* is simply a color phase of *M. mulatta* and should not be considered a different species. The range of *M. assamensis* is from Nepal, through Sikkim, Bhutan, Assam, and upper Burma to perhaps northern Thailand and through Laos to Vietnam. This species is quite consistently found at relatively high altitudes, one variety (*M. assamensis pelops*) ranging mainly between 2,000 and 6,000 feet in the Himalayan region. Very little of its range can be considered lowland and very few sighting reports are from low elevations.

South India and Ceylon harbor two closely related species of macaque, the bonnet macaque (*M. radiata*) and the toque macaque (*M. sinica*), respectively. A third species, the lion-tailed macaque, or wanderoo (*M. silenus*), is confined to the evergreen rain forests of the Western Ghats in southwestern India from about 14° north latitude to Cape Comorin. A rare species, it is probably exclusively tree-dwelling. Its use for an Ayurvedic medicine by the agricultural villagers in Kerala State for a variety of illnesses is threatening it with extinction.

The bonnet and toque macaques are far more common in their respective ranges and are often found on the ground, though not far from the large trees that afford safety. The bonnet macaques are found up to the Godavari River in the east and up to Satara (18° north latitude) in the west. Whether the rhesus and bonnet macaques overlap in the north is still debatable. Bonnet macaques are found throughout southern India, where conditions are favorable, from forests to cultivated areas, towns, and temples. Although they are found in close proximity to lion-tailed macaques in the Western Ghats, typically they occupy bamboo and riverine forests rather than the evergreen rain forests. Webb-Peploe (1947) reported that although *M. radiata* and *M. silenus* live in the same area, the former is not found in the evergreen forests, which seem to be the only habitat of *M. silenus*. My own observations show that bonnet macaques do live in the evergreen forests just west of the Nilagiri Hills.

The toque macaque, *M. sinica,* is found over much of Ceylon, from the dry lowlands in the east to the cool, damp uplands and the low, wet monsoon forest in the west. Today the western part of the range is heavily populated by man, and the monkeys have been considerably reduced in number in that area. The three suggested subspecies (Pocock, 1931; Osman Hill, 1942) have been related by Osman Hill to the three major geological terraces in Ceylon: *M. sinica opisthomelas* to the upper terrace in the wet highlands; *M. sinica aurifrons* to the middle terrace; and *M. sinica sinica* to the drier lowlands.

Southeast Asia has the most varied macaque fauna of any area. Here are found *M. mulatta, M. assamensis, M. speciosa, M. fascicularis,* and *M. nemestrina* in overlapping distributions.

Macaca fascicularis, the crab-eating macaque, lives mainly in the low marshy areas, seaside mangroves, and rubber forests, planted or natural, and in mountain forests along rivers. Perhaps because of its adaptation to these habitats the crab-eating macaque is smaller than the Japanese macaque (*M. fuscata*), and the pig-tailed macaque (*M. nemestrina*). *Macaca fascicularis* is found in Burma, Thailand, Laos, Vietnam, the Malay Peninsula, Sumatra, Java, Bali, Borneo, and the Philippines.

The forest-dwelling pig-tailed macaques (*M. nemestrina*) are found throughout Burma, Thailand, Sumatra, the Malay Peninsula, Borneo and the offshore islands. They have been reported on the Andaman Islands, but they were imported and do not occur there naturally.

Macaca speciosa, the stump-tailed macaque, is found throughout the mainland of southeastern Asia down to 10° north latitude on the Malay Peninsula. Northward they are found up into south China and Tibet where, according to Sanderson (1957), they range the high mountains to the valleys. Thus they may extend up to 12,000 feet in the Himalayas.

Japanese macaques, *Macaca fuscata*, the most thoroughly studied wild macaques, have the most northerly range of any nonhuman primate today. They are found on the major and minor islands of Japan, except Hokkaido, and are scattered over the islands of Honshu, Shikoku, and Kyushu. The island of Yakushima boasts its own local subspecies, *M. fuscata yakui*. The closely related Formosan macaques (*M. cyclopsis*), which are confined to Formosa, differ mainly in having a longer tail than the one-inch stub common to Japanese macaques. Both species are often found in forests that are at least partially coniferous.

The island of Celebes is inhabited by what must be two closely related species of macaques, even though they have been given different generic names. *Macaca maurus* and *Cynopithecus niger* are found on the peninsulas and offshore islands of Celebes.

Baboons and Other Ground-Dwellers   The distribution of baboons (Map 9) is almost entirely in sub-Saharan Africa. Although taxonomists disagree about the number of genera and species (Tappen, 1960), they are treated here as three genera: *Papio, Mandrillus,* and *Theropithecus*. The first two are probably closely related, *Papio* having adapted primarily to the open and the woodland savannahs. Both *Mandrillus* and the gelada (*Theropithecus*) are ground-dwellers, but, whereas the former is a forest monkey, the gelada prefers the safety of regions with precipitous cliffs and immense crags (Crook, 1967). No comprehensive studies have been done on the mandrills.

The five species generally assigned to the genus *Papio* are probably regional variations of a single species rather than distinct, reproductively isolated populations. Four of them—the chacma, olive, yellow, and guinea baboons—are found in Africa from the Cape of Good Hope to Dakar, that is, the open woodland and savannah of Africa south of the Sahara. They are referred to in this book as the savannah baboons; the data given here refer primarily to the East and South African varieties. The chacma baboon (*P. ursinus*) is the southern variety, found in the Republic of South Africa, Southwestern Africa, western Angola, and Botswana. The yellow baboon (*P. cynocephalus*) inhabits parts of Angola, central Rhodesia, Malawi, Mozambique, and the savannah of the southern Congo. The olive baboon (*P. anubis*), which ranges from the savannahs and woodlands of East Africa through the Sudan north of the forest belt in West Africa, is replaced by the guinea baboon (*P. papio*) in the far west.

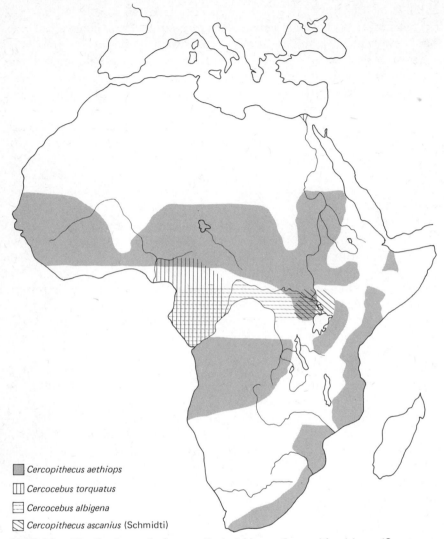

Cercopithecus aethiops

Cercocebus torquatus

Cercocebus albigena

Cercopithecus ascanius (Schmidti)

MAP 10   Distribution of the studied African Cercopithecidae. (*Base map courtesy of Goode's Series of Base Maps, Department of Geography, The University of Chicago. Copyright by The University of Chicago, Department of Geography.*)

The fifth variety, the hamadryas baboon (*P. hamadryas*), is found in parts of the Ethiopian region and in southern Arabia. In the western part of the hamadryas range, there are indications that the savannah baboons and the hamadryas merge (Kummer, 1968). Of the five, the hamadryas baboons are the only ones today living outside Africa.

The gelada, which is probably a special adaptation to seed-eating rather than general foraging (Jolly, 1971), inhabits the western moun-

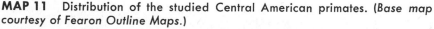

|||| _Ateles geoffroyi_

■ _saimiri oerstedii_

≡ _saguinas oedipus_

MAP 11   Distribution of the studied Central American primates. (_Base map courtesy of Fearon Outline Maps._)

tainous parts of Ethiopia. Despite the fact that it shares both its range and its ground-dwelling proclivities with the hamadryas and savannah baboons, the gelada occupies a different ecological niche.

Another species of Old World monkey that has adapted to ground-dwelling is the patas monkey (_Erythrocebus patas_), which lives in the open savannahs of the northern Sudan and East Africa (Map 9). It relies on speed and diversionary tactics for protection rather than on the large canines and aggressive behavior of the savannah baboons.

Callicebus species

Lagothrix species

Saimiri sciureus

MAP 12   Distribution of the studied South American primates: I. (*Base map courtesy of A. J. Nystrom and Company.*)

Tree-Dwelling Omnivorous–Frugivores   Two African genera, the guenons (*Cercopithecus*) and the mangabeys (*Cercocebus*), are almost exclusively tree-dwellers in the African rain forests. A few, like the green monkey (vervet, grivet) come to the ground for short or long periods of the day, but all remain close to the trees and return to them at the first hint of danger. Their highly arboreal character makes them difficult to find and observe; hence few have been the subjects of even preliminary studies. Map 10 shows the distribution only of those few species referred to in this book. The patas monkey mentioned above is a relative of these tree-dwellers, but partly because of its adaptation, it is classified as a separate genus. The guenons and mangabeys are found abundantly in the rain forests of central and western Africa and wherever gallery forests (forests that grow along rivers and extend only short distances from the water) provide sufficient safety. Since vervets are more ground-dwelling

⊞ *Aotus trivirgatus*

■ *Allouatta* species

MAP 13   Distribution of the studied South American primates: II. (*Base map courtesy of A. J. Nystrom and Company.*)

than most of the others, they are spread widely and adapted to less luxurious forest cover.

Living Prosimians   Only sporadic studies have been conducted on prosimians outside of Madagascar. The African pottos and galagos, the Indian lorises, the Philippine and Celebes tarsiers are all nocturnal and therefore difficult for the bipedal, diurnal ape to observe. In Madagascar, the prosimians have adapted to various terrains, from scrub and thorn forest to rain forest, and are diurnal as well as nocturnal. The genera *Lemur* and *Propithecus,* the best-studied Madagascar prosimians, are spread throughout most forested regions of the island and its offshore islets.

New World Monkeys Like the Old World langurs, colobus, guenons, and mangabeys, the New World monkeys have not been studied much, although the first modern primate field study was done by Carpenter on howler monkeys in Panama. Despite more studies, the free-ranging behavior of most New World monkeys is still a mystery. Only those species that have been sampled are given here.

The nonhuman primates of the New World are all forest-dwelling, living primarily in lowland and montane rain forests and in gallery forests along the major rivers. Most New World monkeys rarely set foot on the ground and hence avoid scrub forests and savannahs. Only in special circumstances has any been reported to descend to the ground; for example, adults will do so to retrieve a fallen infant. Moreover, howler monkeys cross open ground regularly to reach a separate patch of vegetation that provides an abundant supply of desirable food. But the open space cannot be too wide. Under ordinary circumstances, New World monkeys remain in the trees and spend their whole lives far above the ground. There are no ground-adapted monkeys such as macaques and baboons in the New World and no one has yet reported examples of any that are at home on the ground as well as in the trees, such as the vervets of Africa and the Hanuman langurs of India.

New World monkeys live in the tropical rain forests of the Amazon basin and the lowlands of Central America and in the montane rain forests of the Andes, Colombia, Venezuela, and Central America. They do not occupy the savannahs of Patagonia and southern Brazil or the scrub and desert regions of the west coast of South America. Their northern limit is roughly equal to that of the rain forests of southern Mexico (Maps 11, 12, 13; distributions drawn from Napier and Napier, 1967; Hill, 1953, 1955, 1957, 1960, 1962, 1966; and Sanderson, 1957, where not otherwise noted).

Chapter 3
Use of
Space

Animals follow regular and ordered patterns in their daily activities of gathering food, sleeping, playing, mating, and resting. Such regular patterns conserve energy and make it possible for animals to predict the behavior of their conspecifics. There is a disadvantage, of course, since the animal's predators can also predict his behavior and occasionally profit from that predictability by making a meal of him. However, the prey balances this disadvantage by knowing quick routes to safe havens. Moreover, when chased by a predator, the prey often abandons his predictable locomotor habits and resorts to erratic leaping, zigzagging, and bobbing to elude capture (Humphries and Driver, 1967). In general, positive advantages of predictability within the species are so great that few animals rely on random movement or behavior in their daily activity.

Rarely are the individuals of an animal species randomly dispersed throughout their geographical range. As Marler (1968) says, "Individuals may be distributed randomly with respect to each other; they may be evenly distributed, more so than would be expected by chance; or they may have a clumped distribution." Their patterns of dispersion are not unlike those that emerge when human habitations, lots, and farmlands are charted on a map. These patterns are clearly discernible to the other members of the species.

CONCEPTS CONCERNING THE USE OF SPACE

The concept of territoriality has developed slowly over the last century. Altum in 1868 and Moffat in 1903 first presented and then elaborated the idea of territory as feeding space. In 1920, Elliot Howard expanded the concept to include all the space used by a group of animals for various purposes. Actually, despite their emphasis on feeding relationships, both Altum and Moffat recognized that the feeding territory was used for other quite important functions. The current concept, though similar to Howard's, has been considerably refined.

TERRITORY VERSUS HOME RANGE

Burt (1943) clearly defined territory and home range as two distinct entities. Although they have been confused and used indiscriminately, careful study of the actual use of space by different animals reveals two major and distinct patterns. Territories and home ranges are not mutually exclusive: A home range, for example, can include a defended piece of territory. But the two concepts represent different aspects of the way animals use space.

Since Burt's time, *territory* has been defined as that part of the space

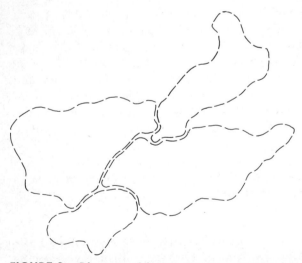

FIGURE 2   Diagram of four territories and their relationships. There is little or no overlap and the boundaries are defended from encroachment by the neighboring groups of the same species.

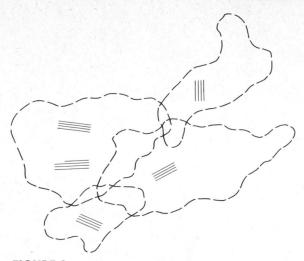

FIGURE 3 Diagram of four home ranges with single-core areas. The overlap of home ranges can be extensive but the core areas are often not overlapped by the range of neighboring groups and are, in fact, defended as territories by some species.

occupied by an animal or group of animals whose area and boundaries are defended from trespass by other members or groups of the same species. On the other hand, Burt defined *home range* as "that area traversed by the individual in its normal activity of food gathering, mating and caring for the young" and as "the area, usually around the homesite, which the animal normally travels in search of food." Jewell's (1966) definition is almost identical: "Home range is the area which the animal normally travels in pursuit of its routine activities." One fact emerges from both definitions: home ranges are not defended and may overlap other home ranges as much as 100 percent.

In order to keep these concepts quite distinct, the term "territory" is limited to mean that space which is actually defended against conspecifics by a group of animals and which is usually marked in some way by them. Whether the defense consists of outright combat or specialized vocalizations, the fact that a space with discernible boundaries is defended makes it a territory. On the other hand, if the animals defend their group from penetration by other groups of the same species but do not actually defend the space per se, the act cannot be considered territorial behavior.

CORE AREAS—FOCI OF ACTIVITY

The optimum habitat for a species of animals is one that fulfills all or most of the requirements of that species in a relatively small space. Sources of food, safe havens for sleep, copulation, play, and social interaction, and enough space to relax undisturbed by outside influences are

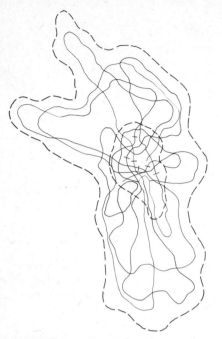

FIGURE 4   The core area is defined by intensity of occupation. A typical plot of ten days' travel by a group of monkeys often shows a concentration of movement within a small part of the total range. Sleeping spots, indicated by short cross lines, tend to be concentrated within the core area.

essential requirements. But only rarely does a group of animals have at its disposal enough optimum habitat that it does not need to move at times into marginal areas.

If the different parts of the space occupied by a group of animals differ markedly, the group is forced to conduct its various activities in different locations within the space. For example, several clusters of trees may be suitable for sleeping purposes, but food must be sought elsewhere. Again, midday resting, grooming, and play may require different parts of the space for optimal benefit to the group. Thus, either because of an uneven distribution of optimum habitat or because of the segregation of activities within the space, the space occupied by a group often shows a wide range of occupancy, from sporadic to regular. Only an exceedingly uniform environment would ensure uniform occupancy. Since such an environment is rare, it is not surprising that the utilization of space by a group varies widely.

Space that is occupied regularly, for whatever reason, has been defined by Kaufman (1962) as the "core area," a useful concept when considering many groups of primates. It consists essentially of marking off spaces that are intensively occupied, as distinct from other spaces that are occupied infrequently, for feeding at certain times of the year when a local crop is available, for resting when the troop has been driven out of the

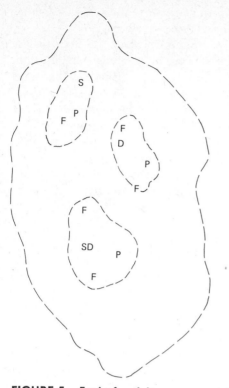

FIGURE 5   Foci of activity are areas of special activity. Although each activity is possible in other parts of the range, the foci of activity are either safer (perhaps more comfortable) for sleeping (S), or have a better food supply (F), or provide the young with a richer play experience (P). If day-resting (D) is done in different locations from night-resting, these may also be foci of activity.

normal core area temporarily, or as alternate pathways from one core area to another.

Though they are optimum habitats in one respect or another, core areas vary in topography and resources. The animals respond to such variation by differentiating their activity in the various parts of the core area. Implying an intense use of the area for the restricted activity, Carpenter's (1940) term "foci of activity" refers to the particular use a group of animals makes of small parts of their range. Monkeys will conduct their activities in that part of the range that is best suited for each activity. For example, if food is available, they will eat in the same tree in which they sleep. However, often the sleeping areas and feeding areas differ because the requirements of each activity contrast strongly: each is used intensively for its own purpose and is a focus for that special activity.

Where core areas and foci of activity exist, they are connected by regular routes over which the animals pass as they move through their home range or territory. For example, the fact that bonnet macaques living in roadside banyan trees move over recognizable paths through the trees is indicated by the bark that has been worn smooth by their daily passage. Other rarely used branches retain their rougher surfaces.

The common pattern for monkeys, including most of the known Old World forms, is for the group to occupy its core area intensively and thus as much as possible to preclude occupation by other groups of the same species. This, of course, is not always possible; if the core areas are scattered, obviously they cannot always be occupied. Exclusion by occupancy is not the same as exclusion by defense. But since many groups defend the integrity of their social group from penetration by other groups, it is often difficult to say when defense of the group changes, almost imperceptibly, into defense of the space the group occupies. Groups of the same species differ in defense or lack of defense of core areas.

MARKING, SPACING, DEFENSE

Primates use three forms of spatial marking: olfactory, vocal–auditory, and visual. The olfactory marks are precise boundary signs that can be easily recognized by other members of the species. They are advantageous because they enable the intruder to recognize his error before the owner is aware of his intrusion. Although vocal calls per se do not indicate territorial behavior, they serve as territorial markers when defense is called for. Otherwise, vocal calls are a spacing mechanism enabling other groups to avoid conflict with the calling group. Visual recognition of boundaries is difficult to demonstrate, but although groups of monkeys can easily see beyond the limits of their range, they remain within circumscribed areas.

Spacing, marking, and defense are interrelated activities. If a territory and its boundary are to be effectively defended, animals must have some means of marking it (Hediger, 1961). If, on the other hand, a group of animals simply occupies a home range and does not defend it, they not only prevent friction by broadcasting their location in some manner but also avoid other groups that may overlap it.

Scent-Marking

Scent-marking is useful for territorial defense since once the marks are left at particular spots, the marker can go on to other parts of the territory. The excellent series by Montagna and his colleagues (Montagna and Machida, 1966; Perkins, Arao, and Uno, 1968) on the skin of primates describes the marking glands of many primates.

Galagos, lorises, and many Madagascar lemurs mark with urine or excrement. For *Propithecus verreauxi* ". . . scent certainly distinguishes a territory, even in the owner's absence, and a territorial dispute involves a frenzy of scent-marking, urination and defecation" (Jolly, 1966). Scent-marks left during normal activity are smelled later by members of another group. Jolly also observed some scent-marking by *Lemur catta* during territorial disputes. Although scent-marking serves other purposes—*Propithecus verreauxi* and *Lemur catta* scent-mark in relation to dominance and sexual behavior as well—it is likely to be used as a territorial marker by groups defending their space.

Wharton (1950b) notes:

The odor of tarsiers is heavy and somewhat disagreeable in a confined place. It clings to trees and wood for days, and is heavy enough to enable natives to spot the general location of a tarsier in its haunts. The writer has been able to detect this odor several times in thick patches of jungle. Some tarsiers in captivity become covered, particularly about the head and hands, with a fine, reddish material which comes off on wood, hands, or cloth. This may be a secretion —certainly it carries a good deal of the odor.

This characteristic of the tarsier, persistent and apparently quite unmistakable, is an ideal territorial marker.

The patterns for scent-marking are varied and elaborate. Jolly (1966) observed "four types of scent-marking in P. verreauxi: throat, urine, fecal and genital with one associated behavior pattern, tail-lashing. L. catta has three sorts of marking: genital- and palmar-marking of branches and marking the tail with the wrist glands. With these three markings occurred two more behavior patterns, anointing the brachial gland with the axillary gland and tail waving." Jolly also reports that other members of the species smell the marked spot with interest or curiosity and that the elaborate gestures of marking visually orient other members to the marked spot.

Sonek (1966) describes the marking of woolly monkeys in a laboratory, and Mason (1968) states that Callicebus have at least two forms of territorial marking, one of which is scent-marking. "Callicebus monkeys seem to have the structural equipment for scent marking (a sternal patch of glandular tissue), and they rub this area against branches by pushing or dragging the body forward."

Jackson and Gartlan (1965) and Gartlan and Brain (1968) report scent-marking in groups of Cercopithecus aethiops both in the wild and in zoos. The scent-marking in the wild occurs at places that other behavior indicates are territorial boundaries. Although scent-marking glands have not yet been demonstrated for vervets, they probably have a mandibular scent gland because of their characteristic marking posture of rubbing cheeks and chin on a particular spot that other vervets later come to sniff. Other means of marking for this species include displays and vocal spacing mechanisms.

Vocal Marking

Vocal marking is best illustrated by howler monkeys and their magnificent howl. Carpenter (1943, 1964) describes the vocalizing produced by the males as "a voluminous roar, low-pitched and ferocious," and that of the female as a "terrible bark." When two howler groups encroach on each other's home range, this vocal marking apparently is a substitute for actual combat. During the regular morning howling session and after feeding or resting, when the group starts to move again, the howling is also used as a spacing mechanism to pinpoint the location of the various groups, which then move in directions that will enable them to avoid contact with one another.

Other primate species also use vocalizations or audible signals to space themselves. The spectacular calls of the Nilagiri langurs, though a poor second to the reverberating roar of the howlers, ensure that the animals rarely come into contact with other groups within the home ranges. On the other hand, the calls of the gibbon initiate territorial conflict by bringing the groups together. The female gibbon gives the "great call," which is followed by a lesser call by the male (Ellefson, 1968) whereas it is the male that issues the warning cry of the Nilagiri langur.

Hall (1965, 1968a) has observed that male baboons in the Urungwe region of southern Rhodesia often bark repeatedly in the early morning while they are still in or near their sleeping trees. This species spacing mechanism of the baboons, similar to the location calls of other species, proves extremely useful in operating home-range systems where the overlap is considerable. If territory is actually defended and the boundaries are maintained, then the contacts between groups occur only or primarily at the boundaries. However, where the home range overlaps as much as 80 to 100 percent among some groups of macaques and baboons, the spacing of the groups becomes a crucial problem; and a spacing mechanism, such as a series of location calls given back and forth, tends to reduce the possibility of conflict resulting from the close contact between groups.

Daubentonia use a grinding call that is answered by other members of the species and may be either a territorial call like that of the gibbons or a spacing call like that of baboons, langurs, and howlers (Petter, 1965). "Mangabeys have several loud vocalizations and it is possible that adjacent groups adjust their movements by these vocalizations" (Chalmers, 1968a). In addition to scent-marking, *Callicebus* use dawn calls in the kind of vocal chain reactions that are particularly effective in group spacing.

The New World marmosets are also reported to vocalize extensively in spacing and defending at least the space occupied at the moment. Thorington (1968), observing groups coming into contact with each other, reported that "with intense vocalizations, animals of troop A chased individuals of troop B approximately 100 yards south." *Saguinas midas*, too, in the early morning sit "on an exposed limb and give a high-pitched vocalization which is answered by neighboring groups and ignored by individuals in the same group" (Thorington, 1968).

Visual–Auditory Marking

A combination visual–auditory signal used by many primates is branch-shaking. When it is done by animals in the trees, it produces a rattling sound audible over a fairly long distance, sometimes a mile or more. Moreover, if the surrounding foliage is not too dense, branch-shaking produces a clearly visible signal. Branch-shaking is commonly done by males. Such varied forms as chimpanzees, gorillas, vervets, baboons, and red spider monkeys reportedly use it as an aggressive display directed against other groups (Goodall, 1965; Reynolds, 1965; Schaller, 1963; Gartlan and Brain, 1968; DeVore and Hall, 1965; Carpenter, 1934).

Sometimes when these animals are on the ground rather than in the trees, they substitute earth-shaking for branch-shaking; here only the visual components of the signal—the jumping, gesturing animal—are effective. Chimpanzees and gorillas have elaborated this mechanism into a spectacular display, tearing off and throwing around foliage, probably to replace the original effect of the branch-shaking that was lost when they descended to the ground. This substitution of branches wrenched from the trees could conceivably be a first step in the use of objects as tools among primates. Gibbons break off far more branches in their aggressive displays than they do in normal locomotion (Ellefson, 1968) and black and white colobus shake the branches by rapid leaping and in the process break off many dead twigs and branches. The falling foliage has some of the effect that gorillas and chimpanzees achieve in their displays. (See Chapter 6, "Communication," for other uses of branch-shaking and associated behavior.)

FACTORS DETERMINING THE AMOUNT OF SPACE USED

The size and form of the space occupied by a group of primates are determined by a combination of factors: the size of the group, the dominance of the group over neighboring groups, the nature and abundance of the food sources, the presence of safe havens for resting, playing, and mating, and climatic alterations. Because of the varied nature of the terrain, probably no two groups of primates solve their space problem in precisely the same way. But similar factors are involved.

GROUP SIZE

Many primates have been studied in areas where several groups live in the same environment. With the environment thus held constant, the groups of different size show the relationship between group size and home range.

One example should suffice. In the wet, deciduous forests at the base of Nilagiri Hills in Tamilnadu, India, two troops of bonnet macaques live along the Moyar River in home ranges that overlap 100 percent. The smaller troop, containing seven monkeys, occupies a range of about a quarter of a square mile, totally within the larger group's range. Twenty-eight monkeys of the larger troop use about one square mile. Both spend much of their time in the riverine forest of bamboo and nondeciduous

TABLE 6. Summary of Group Size and Home Range in Some Species of Monkeys and Apes

Species	Mean Group Size	Group Size Range	Home Range Size (sq. mi.)	Geographical Site	Sources
Howler	18	3–45	0.5	Barro Colorado, Panama Canal Zone	Carpenter (1934)
Gibbon	4	2–6	0.15	North Thailand	Carpenter (1964)
			0.07–0.50	Malaya	Ellefson (1968)
Black and white colobus	13		0.06	Africa	Ullrich (1961)
Callicebus		2–4	1 acre	Columbia	Mason (1968)
Hanuman langur	25	5–150	0.5 –5.0	North India	Jay (1965a)
Chimpanzee	3–5	2–25	6–8	Budongo Forest, Uganda	Reynolds and Reynolds (1965)
			6–30	Gombe Stream Reserve, Tanzania	Goodall (1965)
Bonnet macaque		6–78	0.25–2.0	Mysore and Tamilnadu, India	Simonds (1965)
Gorilla		2–30	8–15	Congo/Uganda Border	Schaller (1965)
Olive baboon	42	12–87	3–16	Nairobi, Kenya Amboseli, Kenya	DeVore and Washburn (1963)
Chacma baboon	34	8–109	3.5 –13	Southern Africa	DeVore and Hall (1965)
Patas monkey	20	5–30	19	Murchison Falls National Park, Africa	Hall (1965)

← Increasing arboreality

Increasing terrestriality →

From Bates, 1970, with slight modifications.

trees and make forays into the deciduous forest away from the river. The ratio of one to four holds for both the difference in troop size and the space occupied.

GROUP DOMINANCE

A dominant group of monkeys tends to occupy the better parts of the available terrain at the expense of the subordinate groups. On Cayo Santiago, the large, dominant group of rhesus monkeys has precedence at feeding stations in the overlapping parts of the home range. Although it may range over the whole land, it has exclusive use of the smaller peninsula on the island and the subordinate groups are limited to the main part of the island. Their range is thus restricted by the pressure of the dominant monkeys. Rhesus macaques in Aligarh, India, show a similar pattern of the dominant group forcing the subordinate groups to move out of favored areas.

The pressure exerted by the dominant group does not necessarily mean that it occupies more space than the subordinate groups. In fact, the reverse may be the case. If the space that is occupied ranges from optimal to marginal, the subordinate groups are forced into marginal habitat much of the time and thus are required to range over more space to meet their requirements. Depending on the amount of optimal habitat available for regular occupation by each group, Struhsacker (1967b) reports marked differences in the size of home ranges for vervets in Amboseli Reserve, Kenya. The subordinate group, whose available optimal habitat was the site of regular intergroup conflict, spent less time there than neighboring groups did in their optimal habitat and more time ranging widely over marginal habitat.

NATURE AND ABUNDANCE OF FOOD SOURCES

Food is often considered to be one of the major factors that limit the occupancy of a particular space by an animal species. Food sources can vary considerably in nature and abundance and in turn can influence the way animals use and occupy space. For example, McNab (1963) distinguished between "hunters" which must search for their food (including fruit, grain, and insect-eaters as well as meat-eaters) and "croppers" (those that browse or graze), to which food is easily available. The difference between them is statistically significant, and the hunters have home ranges four times the size of the croppers' ranges. Though primates eat insects and fruit and some occasionally eat meat, their diet includes the browsing and grazing habit of croppers and their home ranges tend to be considerably smaller than those of true carnivores. Man is the only higher primate habitual hunter and man's home range is many times the size of the cropping baboons (Washburn and DeVore, 1963). Human hunting bands may use 100 square miles and more, while baboon troops rarely exceed 15 to 20 square miles.

Biomass and Carrying Capacity

A way of considering the relationship between food and space is through the concepts of biomass and carrying capacity. The biomass is the sum of the individual body weights of a group of animals living in a delimited area and is usually expressed in terms of the total weight of the animals per unit of space (for example, kilograms per square kilometer). The carrying capacity of an area is the total amount of living matter that it can support and depends not only on the mineral and climatic resources but also on the particular mix of animals and plants that inhabit the area at one time. A climax forest supports a different fauna than a regenerating forest on the same land. The carrying capacity of an area can be calculated by adding the biomasses of the species living in the area. Since the carrying capacity of the terrain influences the size of a group's range, changes in carrying capacity are important. In East Africa the elephant population has been expanding in recent years. Gartlan and Brain (1968) describe some results.

The variations in the vegetation seen at Chobi are brought about by the grazing pressures of the numerous game animals. The principal diet component of elephants is grass and herbs, but they cause considerable damage to tree species. *Terminalia glaucescens,* a main diet component of monkeys, is very susceptible to elephant damage, and dies when barked; it is also easily uprooted. Large areas of the Murchison Falls National Park that were once dominated by this tree are now, to all intents and purposes, treeless *Hyparrhenia* grassland.

In this marginal habitat the vervets must not only move over much wider areas to find enough to eat but also spread out in their search much more than vervets living in richer environments.

Territoriality and home-range occupation are means whereby population density can be controlled, but density itself is a complex factor that involves more than just the numbers of individuals living in an area. The concept of biomass helps to explain the additional aspects of density by relating the energy requirements of the species living in an area to the carrying capacity of the area. Thus an important part of population density, and thereby the amount of space used, is the amount of body weight that must be maintained for a particular species in a given area.

But animals vary in their metabolic rates as well as in size. For example, many shrews must eat their own weight or more in food each day to maintain their high level of metabolic activity, whereas many of the larger and more advanced mammals can subsist on a relatively smaller amount of food and still maintain their regular metabolic activity. As a result, in many cases even the biomass is not so fine a measure as could be desired, and metabolic energy expenditure more closely approximates the actual support required by a particular group. However, McNab (1963) has shown that biomass and the metabolic function are not significantly different for many purposes.

Not only the amount but also the intraspecies distribution of the biomass is an important factor to consider. Sexual dimorphism, for example, affects the distribution of the biomass, the larger males requiring more support from the local environment than the females. Washburn and DeVore (1963) argue that where heavy predation prevails, the larger males are necessary to fight off predators. On the other hand, females, the producers of offspring, are not nearly so large. The sexual dimorphism in size allows an efficient use of resources while combining a high reproductive rate with large protective males. Lacking the heavy predator pressure of terrestrial species, the arboreal gibbons also lack sexual dimorphism in size.

HAVENS

Food is often considered to be one of the major factors that limit the occupancy of a particular space by an animal species. However, other factors sometimes take precedence. For example, sea birds that nest on cliffs are limited to nesting on those islands which have shore lines with steep cliffs or other suitable features that offer adequate protection. Since they are free to move out over wide areas of the sea in search of food, their territoriality is concerned primarily with establishing a nesting place, not a feeding area.

For most primates, the major problems of survival are feeding and securing havens from predation. The latter differ somewhat from the nesting sites of sea birds, since the mother primate takes her infant with her wherever she goes. However, some activities require a time and a place that are free from the pressures of predation, from other groups of the same species, and from other animals pursuing a similar way of life. Reproduction, in fact, necessitates uninterrupted time for the male and female to negotiate a mating and to copulate in safety. Individuals involved in copulatory behavior are highly susceptible to predation and often will not attempt to mate if there is any uncertainty about their surroundings. Jane Van Lawick-Goodall reports that she did not begin to see mating behavior until she had been with the chimpanzees for nearly two years, that is, long enough for the animals to relax sufficiently in her presence to copulate.

Fortunately, most primates are tree dwellers and thus have innumerable safe hiding places, and the determining factor in setting up the boundaries for the group space is primarily an adequate source of food. But some species—hamadryas baboons, savannah baboons, patas monkeys—live in areas where safe refuges are restricted (semidesert, open savannahs); for such, the sources of food are often of secondary importance. Baboons require within their home range night resting places that will adequately protect them from predation; lacking them, no baboons will inhabit the area regardless of a plentiful food supply.

Bonnet macaques, which are basically forest-dwelling monkeys, have in recent times adapted to living in towns and along roadsides. In years of normal rainfall, these monkeys find plenty of food in the fields between the villages. Their primary requirement, which limits their distribution in southern India, is to find large trees to which they can return when danger threatens and in which they can sleep safely at night. Sometimes tall buildings or steep hillsides provide suitable substitutes.

The major limiting factor in the size of the annual home range or lifetime range of gelada baboons is the presence or absence of cliff gaps. Any appreciable gap in the cliffs blocks the group from moving across to new areas. Otherwise the gelada herd occupies as much area as it can close to a continuous or relatively continuous cliff edge that ensures their sleeping safety (Crook, 1966). Seasonal differences also affect the size of gelada herds and thus the area they traverse. In April, when they tend to be smaller, the largest herd is around three hundred and the smallest twenty-five; in February the largest reaches four hundred, the smallest thirty. Crook points out that after the rains large herds forage for food among the crops in agricultural areas and remain around the villages during harvest. Later, however, as the food supply decreases, the animals are forced off the agricultural land and the herd size decreases. The local movements "involve shifts into gorge sides or into areas where grasslands are more extensive" (Crook, 1966).

Another indication of the importance of safe havens is that refuges and sleeping places are generally found within the core areas. Females with newborn infants tend to remain closest to the core areas since they are the safest part of the home range. "As one traverses the [gelada] herd away from the cliff edge the numbers of infants and babies rapidly decrease while that of adult and large subadult males increases until, on the outside flank, only males are observed" (Crook, 1966). The clustering of females with young infants has been reported for many other species (chimpanzees, Kortlandt, 1962; baboons, DeVore and Hall, 1965) and, at least for bonnet macaques, mothers tend, like the gelada, to stay closer to the safe havens than do the males in the troop.

Thus Jewell's (1966) definition of occupied space, "Fundamentally the home range is an area with a certain productivity that meets the energy requirements of the group or individual that occupies it," is incomplete since it does not include the very important element of safe havens.

SEASONAL CHANGES AND CLIMATIC CYCLES

Various reasons can induce an animal or group of animals to shift to a new range. Climatic cycles, which make certain foods available at different times of the year, are one of the main causes for such shifts. Unless barred from new food resources by excessive predator pressure or the occupancy of the area by another group of the same species, the animals will accommodate to the new source. The patterns of group movements often show an annual cycle with seasonal fluctuations. For example, when the vegetation is lushest and food is most plentiful, baboons are not likely to move far from their sleeping trees (DeVore and Hall, 1965; Hall, 1962a). Because rhesus monkeys in Uttar Pradesh, India, make only seasonal use of sheesham trees, they are found in forest dominated by sheesham trees from January to March but they apparently shift their range to a more mixed forest for the rest of the year (Neville, 1968).

Long-term climatic cycles change the availability of food and refuge areas even more markedly. Cycles of vegetational change of more than a year are usually reflected in a more permanent redistribution of space among the groups of animals occupying the area (an example is given in Chapter 4, page 107).

PRIMATE HOME RANGES

Under normal conditions most of the wild primates about which we have information are not defenders of territories. Territorial defense is, after all, an expensive activity in terms of the efficient use of energy. To be effective, the policing of territorial boundaries must be regular and systematic; that is, the group occupying the territory must be constantly aware of the potential threat to their occupancy by other groups on the periphery. On the other hand, without territorial defense, the interactions and therefore the overlap between groups of the same species are greater; the area to be exploited is larger; and the single group, relieved of the need for regular policing, is free to move more randomly through the home range, maintaining their familiarity with it and keeping track of the approaching ripeness of various fruits and the availabiilty of other sources of food. In home-range systems, adequate spacing mechanisms maintain the distance between the troops and reduce the possibility of conflict. Neighboring groups are repelled rather than attracted by signals indicating the presence of others.

The various macaque species occupy home ranges that often overlap considerably with those of neighboring groups of the same species. Since two groups do not occupy the same space at the same time, conflict can occur when two groups approach each other. But no one so far has reported that they regularly defend the boundaries of their space. Under the stress of overpopulation (for example, rhesus monkeys living in a temple compound—Southwick, Beg and Siddiqi, 1965), aggression rises markedly and conflict occurs whenever the larger, more dominant group contacts the smaller groups. However, the latter are driven not from the space occupied by the larger group but only from the immediate vicinity. On Cayo Santiago, the smaller groups live within the range of the dominant group, which has free access to all parts of the island. The overlap is total, negating the exclusive use and defense of territory as a behavior pattern for these rhesus macaques.

Among baboon groups, fighting is exceedingly rare, perhaps because in many cases the relationships between groups have been established for a long time. Each knows what to expect of the other. Therefore a few warning gestures or a slow displacement usually suffices to prevent the two groups from simultaneously occupying overlapping home ranges. When, however, groups of monkeys are forced by changes in vegetation (natural or man-made) or other shifts in local ecology to come into contact without sufficient time to make the adjustments that would lead to compromise relationships, a period of active conflict sometimes precedes the low-key interaction of more established groups. DeVore and Hall (1965), for example, describe fighting among groups trying to settle in the same trees in Amboseli. A new troop of bonnet macaques, which had moved into the region of the Somanathapur troop near Bandipur because large tracts of bamboo in its old range had gone to seed and died, was not treated with the same detachment accorded to older, established neighboring troops to the north.

Hamadryas baboons (Kummer, 1968) manifested no cases of territorial defense. Their penetration of one another's home range is virtually complete, including even the sleeping cliffs. Having definite differentiation of function within their home ranges, they did not feed (an activity that can easily lead to conflict) near the sleeping cliffs although food was available there, nor did they sleep, copulate, or fight en route to and from the sleeping places (Kummer, 1968).

Most gelada herds probably move less than 4½ miles a day in contrast to the greater distances covered by hamadryas and savannah baboons (about 12 miles). Their utilization of their home range is very different from that of savannah baboons and macaques. Their herd structure is such that it changes from day to day; only the one-male units seem to maintain their integrity (see Chapters 8 and 9). Since the herd consists of numerous one-male units and the herd changes composition from day to day as one-male units are added or deleted, the area that is occupied to support a particular herd also changes; in fact, it would be difficult to distinguish a home range occupied by one group of gelada from that occupied by other groups, particularly at the one-male group level. Instead, the lifetime range or annual range of the whole herd is probably more significant, although it is not directly comparable to the range of single troops of savannah baboons or macaques.

Whether there are definable home ranges for gorillas is questionable, since they move through wide areas and overlap extensively with other groups. Occasionally two gorilla groups will join together and move for some time as a unit before separating; in this respect they are far more flexible than baboons. Chimpanzees are also less bound to a specific area for a specific subgroup, although the total chimpanzee population of a region may well have definable limits for its home range. The large-sized groups described by Itani and Suzuki (1967) are usually distributed at random in the forest patches of Uganda. When they move from one forest patch to another, all move together, leaving one forest patch with no chimpanzees. Since they move from patch to patch over open and dangerous ground, they probably know what they are moving into; the forest patches correspond to core areas, and the open ground is used primarily for travel routes.

PRIMATE CORE AREAS

Bonnet macaques living along the roads in Mysore State usually occupy the tall banyan trees lining the roads. Although they forage for food, most of their time is spent in and around the banyan trees, which thus qualify as core areas. Certain of these trees are more desirable than others and hence are occupied more frequently, either because they have better fruit, because they form particularly good play areas, or because they are suitably safe and comfortable resting places.

Along the highways of Mysore State, double rows of banyan trees form single, fairly uniform core areas for bonnet macaque home ranges; but even within the rows, foci of activity can be distinguished. Generally a play spot is near adult resting spots but distinct from feeding areas.

In some of the forested regions of southern India where the vegetation is mixed, the bonnet macaques find certain flora, such as large bamboos, especially safe for sleeping. Since the bamboo is scattered, the macaques usually sleep in the same clump night after night. For baboons, also, which move out onto the savannah to eat during the day but return to safe resting and sleeping places at night, their activity is concentrated in special, isolated parts of their total range.

Groups of the same species that occupy more uniform regions (that is, areas where no demarcated zones are either safer or more desirable than other parts) do not focus their activities on special parts in such a way as to produce core areas. In some parts of their range, bamboo-dwelling bonnet macaques occupy large tracts of mostly optimum habitat, that is, stands almost totally bamboo. Unlike bonnet macaques in a mixed forest, they do not return to the same clump night after night but move to new sleeping places, almost at random (Nolte, 1955).

As another example, except for a predilection for certain types of vegetation, *Propithecus verreauxi verreauxi* do not focus their activity in core areas (Jolly, 1966). Since these prosimians must have open stems to leap to as they move through their home range, they avoid areas that are characterized by low, bushy foliage. Thus, within the boundaries of their home range and territory, they habitually occupy certain regions and avoid others, but the occupied areas are used about equally.

Thus the mixed forest-dwelling or roadside troops of bonnet macaques and the savannah dwelling baboons, which live under conditions that require strong differential use of space, tend to have distinct core areas or foci of activity. On the other hand, the bonnet macaques observed by Angela Nolte in a region of forest predominantly bamboo showed a lack of core areas for sleeping activity. *Propithecus verreauxi verreauxi* represents an intermediate condition where the only differential use of space was a strong tendency to avoid areas unsuitable for locomotion. So uniformity or lack of uniformity in the environment is probably a much stronger determinant of the incidence of core areas than is the species itself.

Concerning defense of core areas, Jay (1965) reports that among the groups of Hanuman langurs there is no overlap of core areas. The home range outside the core area sometimes overlaps. Yoshiba (1968) reports that Dharwar troops of Hanuman langurs actually defend their core areas.

Vervets also defend the optimal habitat that forms the core areas of their ranges (Struhsacker, 1967b). Of three groups he observed, the larger two spent 95 percent of their time in optimal habitat, whereas the smallest spent only 68 percent of its time in the optimal habitat; the rest of the time it had to forage over a wider area of marginal habitat than the first two groups. They may have avoided the optimal habitat because of the frequent intergroup aggression occurring there and may have been forced to utilize the larger, but less productive, area.

Although roadside bonnet macaque troops sometimes briefly occupy the core areas of other troops, they do not use these areas as they would their own even though the ecology is identical. Usually the invading group feeds for a brief time and departs when the owner troop returns.

A core area can be so habitually occupied by a single group that other groups do not have the opportunity to use it. Such a core area is defined simply by use and does not involve active defense by the occupying group. On the other hand, it is possible to have the intensively used portions of the home range defended in some manner by the occupying group while the rest of the home range goes undefended. Both the Hanuman langurs observed by Yoshiba and the vervets observed by Struhsacker exemplify this kind of defense. The marking of boundaries is the usual criterion for this defense, and langurs do it vocally. Thus a defended core area is really a territory within a home range. Many animals range farther than the limits of their defended territories.

PRIMATE TERRITORIES

Although the modal pattern of behavior is often either territorial or restricted to home-range occupancy, primates do not neatly divide into territorial and home-range species. There are examples of some groups of a species using territorial defense, while others do not. Jay (1965) reports Hanuman langurs inhabiting home ranges which they do not defend. On the other hand, according to Yoshiba (1968), the Hanuman langurs of Dharwar have a fixed home range and a core area which they defend as a territory against other groups of langurs. The peripheral areas of the home range outside the core area considerably overlap the ranges of other troops. Jay's information indicates that core areas are not

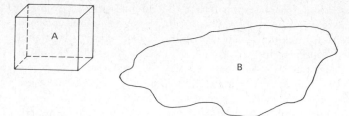

FIGURE 6   Group A's defense of its three-dimensional space is easy in contrast to any attempt by group B to defend all of its two-dimensional space.

used jointly by neighboring groups; those she observed may have made a better adjustment to the neighboring troops and no longer showed any overt aggression. In any event, one cannot flatly say that Hanuman langurs are either strictly territorial or lack defense of their space. This may be true of other species for which we have information only from a few groups.

More studies of primates in the wild show that increasing numbers of primate species do defend a territory. Although active defense of the boundaries of the occupied space may not be the most common use of space, it is no longer as uncommon as was thought a few years ago.

Among most animals, defense of a territory is not part of a pattern of aggrandizement or expansion but rather a means of maintaining a stable relationship with other members of the same species in the same general area and with the local ecology. Because of the confrontations in defense of territory, the territorial spaces occupied by the various groups are continually reinforced and the balance is maintained.

Territorial defense requires that the two groups of primates concerned must actually come into contact with each other at the boundaries and assert their rights. Highly arboreal species, which occupy compact spaces having not only length and width but also height, are more efficient in defending territories than the usually larger groups of ground-dwelling primates, which move over wide areas of two-dimensional space. The tree-dwelling forms can fairly quickly reach any part of their range, whereas a baboon troop is often several miles from parts of its range. Those species that are reported to be characteristically territorial, *Hylobates lar* (gibbon), *Saguinas midas* (tamarins), and *Callicebus moloch ornatus* (titi monkey), live as mated pairs with their offspring (see Chapter 7, "Social Organization," and following chapters) in small territories. Their territorial defense and their small social groups are probably interrelated and contrast with the troop organization of many nonterritorial primates. The territorial Dharwar langurs also have small bisexual groups, unlike the nonterritorial troops Jay observed.

Two or more gibbon groups meet at the boundaries of their territories and, through a series of displays leading sometimes to actual fighting, defend those boundaries.

Prolonged conflict-hooing by the adult males (33 minutes on the average) precedes the first chase of an encounter. This prechase period is foreshortened when groups meet in a situation directly involving a food source; it is protracted when a food source is not involved. . . . During the initial conflict-hooing session, the males put on an acrobatic display that includes twisting and dangling around on one arm and swinging rapidly around a particular arboreal pathway (usually a circular or oval route of from 10 to 20 yards). . . . In 14 percent of the intergroup conflicts large dead branches are broken off by the adult males during their displays and subsequent chases. The branches crash to the forest floor (Ellefson, 1968).

These are obviously displays rather than errors in locomotor judgment since "it is extremely rare that an adult gibbon mistakes a dead branch for one that will support it" (Ellefson, 1968). The conflict is usually a chase.

As far as we know, *Callicebus moloch* and *Saguinas midas* are the only proven examples of territoriality in the New World. It is unlikely that the howlers, *Allouata palliata,* are territorial in the strict sense since their home ranges overlap considerably and the roars they emit in the morning or at sight of other groups of howler monkeys do not attract them to each other but usually cause them to disperse.

The group structure of *Saguinas midas* and *Callicebus moloch,* a mated pair of adults with their offspring, is similar to that of the gibbons. Mason (1968b) reports that in the area of his study *Callicebus* territories overlap from 5 to 7 percent at the most. Mason contends that *Callicebus* displays true territorial behavior in the sense of defending its space against encroachment, unlike those species (for example, baboons, macaques, howlers) whose overlap between home ranges is as high as 60 to 100 percent. *Callicebus* territorial displays are described by Mason as follows. In the small overlapping areas at the boundaries of the territories,

peer groups met frequently and engaged in an elaborate and extended display in which calling, rushing, and chasing were prominent elements. Such intergroup confrontations characteristically began with two of the groups rapidly converging toward each other, and stopping a few feet apart. As the groups approached each other the animals of one group would draw together, and then sit with their sides touching, facing the opposite group. Calling was probably most often initiated by the male but his mate certainly participated, and both animals showed signs of extreme agitation. . . . Vocalizations were loud and sustained. Piloerection, arching of the back, stiffening or bowing of the arms, and tail lashing frequently accompanied the calling, and the rushes, counter-rushes, and chases that normally occurred during the confrontation were carried out with great vigor (Mason, 1968b).

Such displays usually serve to reinforce the boundaries and dispense with the need for actual fighting. Occasionally a chase occurs, but pursuit seldom results in physical attack. The lack of wounds reported by Mason is an interesting indication that fighting is exceedingly rare. The calling could easily function as it does among gibbons, to bring on a confrontation and thus to reinforce the boundaries of the territories.

The *S. midas* tamarins apparently show territorial behavior among groups. According to Thorington (1968), groups came into contact with each other and "with intense vocalizations animals of troop A chased individuals of troop B approximately 100 yards south." The chases occurred between the normal ranges of different pairs, that is, along the boundaries where there is little overlap of range. If this is the case, then these tamarins show territorial patterns very similar to those of *Callicebus* and *Hylobates*. The small, mated-pair group is probably an efficient way of spacing a species, and territorial defense between such pairs is likely to be fairly evenly matched. Thus confrontations can be handled with displays and chases, with little danger of either degenerating into a fatal fight. As a society increases in size, the matching between two neighboring societies is less likely to be equal, and the territorial conflicts could degenerate into brawls that would result in several deaths. If such conflicts occurred regularly, the population would be rapidly decimated.

There is some evidence that both *Propithecus verreauxi* and *Lemur macaco* defend territories. J. J. Petter (1965), for example, observed some fighting between two groups of *Lemur macaco*. Jolly (1966) says that *Propithecus verreauxi* have territories that are within their home ranges. The territories are not entered by other groups. The home ranges overlap considerably and territorial disputes occur in the overlapping space. *Lemur catta* also have this pattern of overlapping home ranges containing central territories that are defended against conspecifics (Jolly, 1966). Territorial defense is probably common among many prosimians; the marking glands on the forearms and under the chins of these species are best explained as territorial marking glands.

INTERSPECIFIC CONTACTS

Several different species of closely related animals can and do occupy the same general area, but only if they do not actively compete with one another. Among primates, there are numerous examples of different species living in fairly close association, even occasionally intermingling, without any major conflict and with limited competition. In India it is fairly common to see langurs and macaques feeding together in the same trees at the same time. Conflict arises only from occasional displays of dominance by the macaques. Occasionally, in fact, the two species are found in bispecific (or bigeneric) groups; Phyllis Jay's Kaukori troop of Hanuman langurs included a pair of rhesus macaques well integrated into the social group of langurs.

Groups of different species associating in this manner and occupying similar or identical ecological niches sympatrically often develop a dominance-ranking among themselves. When they come into contact over the same food sources, macaques generally dominate langurs but rarely carry the dominance to the extreme of excluding the others from the food even when both are exploiting it at the same time. In much the same way, the more ground-dwelling *Lemur catta* in Madagascar dominate the less aggressive and highly arboreal *Propithecus verreauxi*.

Whereas each species' use of space is a separate system (defense of territory is generally directed only toward members of the same species), the relationships among groups of different species are probably not established without some friction. Most of the contacts so far described are among groups that have had a history of contacts before the observations began. One example of a new interspecific contact will show that some friction is likely at first.

In early 1962, a small group of Hanuman langurs moved into the home range of the Somanathapur troop of bonnet macaques that had been observed for several months. Each time the langurs came close, they were threatened vocally and with gestures by the bonnet macaques and were actually chased away several times. In fully established relations between the two species, such threats and chases are rare.

In any event, the langurs almost always give way to the macaques, even though they are twice the size of the latter. But the animals feed together in the same trees successfully because the langurs are leaf-eaters and the macaques cannot digest mature leaves. Both can eat new shoots, but since these are not the sole source of food for either species and their other sources differ, they do not compete markedly with one another. Because the difference in their total diets is such that an occasional exploitation of the same food source is not detrimental to either group, they continue to live amicably side by side.

Crook and Aldrich-Blake (1968) describe a situation in which doguera (savannah) baboons and gelada live in the same forested region without real competition because their exploitation of the environment differs. The doguera baboons feed primarily in the forest, concentrating on opuntia leaves and fruit, which apparently the gelada never touch. To be edible, these prickly pears require delicate handling, a technique that has been mastered by the doguera baboons but not by the gelada. The latter move out into the open country and spend a long time eating grass; since grass has a lower food value, greater quantities must be obtained and more time must be spent gathering it. Moreover, doguera baboons readily climb into the trees, whereas the gelada are more ground-dwelling, moving into the trees only rarely. During the rainy season, large groups of gelada forage together, including all-male groups that mingle in the herds with a number of harems. During other seasons, the smaller harem groups forage on their own.

Gelada and doguera baboons have been observed to intermingle briefly with little interspecific conflict. What conflict there was involved only threats and lunging, not actual attacks. Usually they pass within a few feet of each other, apparently taking no notice of one another, much like the macaques and langurs that eat together in the trees of India and Ceylon.

When frightened, the gelada make no use of brush or tree cover but bolt directly for the cliffs. Like macaques in the forests, they move across the plains silently, disappearing quickly when startled. The doguera baboons are more likely to take refuge in the trees, when necessary.

Contacts between hamadryas and gelada are also relatively uneventful, with only the smaller animals avoiding the larger ones of the other species and the females avoiding the males. Here again, the differences in feeding are fairly marked: the hamadryas with their long legs and long stride move over large, open spaces of countryside, covering great distances and exploiting sparser conditions than can be reached by the gelada, which tend to sit to eat and to move over relatively short distances.

Chapter 4
Cycles
of Activity

The life cycle of any animal form includes the phases of its existence from conception to death from natural causes. The generalized sequence—prenatal life, infancy, youth, maturity, and old age—characterizes numerous life forms.

THE PRIMATE LIFE CYCLE

In the primate order, which includes a wide variety of animals from the primitive tree shrew to man, the specific life cycles vary from species to species. But an overall pattern, which is shared with other orders of mammals, characterizes them all, and the differences between species are often primarily differences of duration. There is a general, but by no means universal, tendency for the smaller and more primitive members of the order to have a shorter life span and for each phase within their life span to be proportionately reduced. Nevertheless, one can as accurately speak of the infancy, maturity, and old age of a tree shrew as of a man. In fact, the growth pattern and aging process among tree shrews are so much more similar to those of man than to those of rodents that attempts are being made to develop a strain of tree shrews suitable for laboratory studies of these processes.

THE PRENATAL PHASE

The prenatal phase of the life cycle, which has been carefully analyzed in man, includes a series of stages from conception to the end of the second month, when the metamorphosing embryo becomes a fetus. In the first sixty days, the human embryo develops from a single fertilized cell into a recognizable human fetus. The nervous system is formed, the muscular system develops, and the new individual takes on the form of a recognizably distinct human being.

The fetal stage is primarily one of growth, although there are detectable changes in form. The cells of the fetus are expanding mainly in size rather than in number, a process that continues after birth. Except for some details and its inability to survive outside the womb, the fetus is a complete individual. Although the prenatal development of other species of primates is not well known, what is known generally conforms to the pattern of prenatal human life.

The period of human gestation usually lasts about 267 days (*Biology Data Book,* 1964). The various apes—gorilla, chimpanzee, and orangutan —often have somewhat shorter periods although they overlap the range established for man (Table 7). Among the Hominoidea, the smallest ape, the gibbon, has the shortest period of gestation. Old World monkeys, including macaques, baboons, langurs, and various other cercopithecoids, usually have gestational periods of about six months; among the New World monkeys, the periods are generally closer to four to five months. Gestational periods of four months are normal for the large prosimians, although variation is indicated by the five-month period reported for *Propithecus* (Table 7) (Petter-Rousseaux, 1964) and the two-month period for the smaller forms like *Microcebus* and *Chirogaleus* from Madagascar (Petter-Rousseaux, 1964) and the more primitive forms like the Philippine tree shrew *Urogale* (Wharton, 1950a).

Size, level of evolutionary development, and length of gestation seem to be closely correlated. The more advanced and larger forms have longer periods of gestation. Why then does man, whose evolutionary advancement is so much greater than that of the other great apes in terms of environmental control and ability to adapt through culture rather than through genetic and subsequent phenotypic change, not have a proportionately longer period of gestation? In terms of its motor and sensory development, the human infant is born earlier than the other great apes because of the rapid growth prenatally and postnatally of his brain: it must be born before the brain and head are too large to pass through the birth canal. Thus, if the human infant were to be born with motor coordination as great as a newborn ape's or monkey's, the period of human gestation would be longer by many weeks.

TABLE 7.   Periods of Gestation in the Primates

Species	50	75	100	125	150	175	200	225	250	275	300	325
Apes												
Homo sapiens									XXXXXXXXXXXX			
Pan gorilla									XXXXXXXXX			
satyrus								XXXXXXXXXXXXX				
Pongo pygmaeus									x			
Hylobates lar							x					
Old World Monkeys												
Papio hamadryas						XXXXXX						
ursinus (etc.)						XXXX						
Erythrocebus patas					x							
Cercopithecus aethiops						x						
Presbytis entellus						x						
Macaca mulatta					XXXXXXXXXXX							
sylvanus							x					
radiata					XXXX							
nemestrina					x		x					
fascicularis					xxx		x					
New World Monkeys												
Ateles ater				x								
Lagothrix humboldti				x								
Alouatta seniculus				x								
Saimiri sciureus					x							
Leontocebus rosalia			xx									
Prosimians												
Propithecus verreauxi					x							
Galago senegalensis				x								
Lepilemur mustelinus				XXXXXXXX								
Lemur macaco				XXXXXXX								
fulvus				XXXX								
catta				XXXX								
Cheirogaleus major		x										
Microcebus murinus	xx											
Urogale everetti		x										

POSTNATAL LIFE

Postnatal life is commonly divided into infancy; childhood, or juvenile period; adolescence, or subadulthood; maturity, or adulthood; and finally, old age. In field studies of primates, these major stages of postnatal life are often subdivided into smaller units. Jay (1963) and DeVore (1963) both use a fivefold division of infancy. Schaller (1963) does not subdivide and talks only of infants in general. Others (Simonds, 1965; Carpenter, 1965) have used divisions simpler than Jay's and DeVore's, more complex than Schaller's. The differences between these categories are due mainly to differences in emphasis or to the amount of data collected about infancy. The phenomena of growth and development form a continuity in which more or less sharp transitions are interspersed with many that are

TABLE 8.   Infant and Adult Coloration in Several Species of Primates

Species	Infant Color and Duration	Adult Color
Various macaques	Dark; 2 months	Red-brown to grey-brown
Savannah baboons	Dark; 4–6 months	Olive-brown
Hanuman langur	Dark fur, pink skin; 3–5 months	Grey fur, black face
Mountain gorilla	Skin pinkish grey to light or medium brown; 1½ to 2 months	Dark brown to black
Lemur macaco	Dark; 6 months at longest	♂ dull black, ♀ very light red-brown
Howler monkey	Greyish; 5–6 months (Adult mantle begins to appear in second year)	Black with reddish mantle
Chimpanzee	Skin pale pinky-yellow; 1–2 years	Darker
Red spider monkey	Black; about 6–10 months	Reddish
Gibbons	No color distinction	
Various *Cercopithecus*	Natal coat; 2–4 months	Species adult coat
Allenopithecus negroviridis	Lighter than adult; 10 weeks	Dark
Red and olive colobus	No color distinction	
Black and white colobus	Distinct natal coat	Black and white

blurred. Differences of development among species and genera further complicate the categorizing. In general, the more detailed the study, the greater the number of transitions.

Infancy

Infancy is a period characterized by extraordinary dependency upon the mother immediately after birth and by a close contact between mother and offspring that affords protection and sustenance to the neonate. In many species of primates, infancy can be divided into two stages based on color phases. The newborn infant, which is often a different color from the adult (Table 8), changes to the adult coloration sometime within the first few months. Since this change is easily observed, it is the basis for defining one stage of infancy in most of the subdivisions.

In macaques of all species, the dark brown to almost black coat of the infant contrasts with the red-brown to grey-brown of the adults. The same sharp contrast characterizes the savannah baboon, but the natal coat color persists until the fourth to the sixth month; macaques generally change color in the second month. According to Cynthia Booth (1962), all *Cerco-*

PLATE 8   The Nilagiri langur (*Presbytis johnii*). This male infant is about six months old and has the adult color.

pithecus monkeys in Africa show the contrast between natal and adult coat color. Among the leaf-eating monkeys, the coloration is more complex. A. H. Booth (1957) noted that among the red and grey *Colobus* monkeys of Africa, the infant and adult coats are similar in color, whereas among the black and white *Colobus* and the Asian langurs the color contrast is marked. The common Indian langur is a study in contrasts; he has not only dark natal hair that contrasts with the light grey of the adult but also a pink face that contrasts with the black face of the adult (Jay, 1963, 1965). These langurs change to the adult pelage and skin color between the third and fifth months (Table 8).

Among the New World monkeys, the same marked contrast between natal and adult coat color was reported for howler and spider monkeys (Carpenter, 1934, 1935). Many other New World species undoubtedly show this same contrast. Petter (1965) reported the natal coat color for *Lemur macaco* in Madagascar but did not observe the time of the change to adult coloration, only that it had taken place by the end of six months.

Gibbons apparently do not show a contrasting natal coat color (Carpenter, 1940), but at birth chimpanzees and gorillas have less pigment in their skins than the adults. Unlike the mountain gorillas whose pigment forms quickly (Schaller, 1963), chimpanzees wait two years for the pigment to attain the adult coloration (Reynolds, 1965; Goodall, 1965). Chimpanzee infants and juveniles also have white tail tufts, which generally disappear by young adulthood.

The period of early infancy, which in many species is marked off from later infancy by a color change, has its defining behavioral characteristics as well. For the most part, these will be discussed in detail in the chapter on adult–infant relationships; during infancy they are basically linked to the exceedingly close contact between mother and infant. The newborn infant, especially in the first few weeks (which stretch to months among chimpanzees and gorillas and up to a year or more in man), is totally dependent for all its needs upon its mother, who throughout the period of early infancy is exceedingly solicitous.

Later Infancy

The transition from early to late infancy in many species is marked not only by the color change but also by a gradual loosening of ties between mother and infant. The period of late infancy then includes the time from the color change to the infant's rejection by its mother. In many species maternal rejection occurs when the mother gives birth to her next infant; but in others (for example, savannah baboons), the infant is rejected when the mother reenters the reproductive cycle with her first postpartum estrus.

Later infancy can be divided into three parts: (1) the period of the color change; (2) the period after the color change in which the infant is still allowed to nurse; (3) the weaning period, which in many species often stretches to the birth of the new infant. In macaques and other monkeys, which normally give birth every year, weaning is often not completed before the mother's next pregnancy and is sometimes extended until almost the birth of the new infant. Lactation probably does not occur after the mother reenters the estrous cycle, but the infant is still carried and permitted to suckle the nipple for comfort. It is more like the security-seeking thumb-sucking or blanket-clutching of human infants than a means of nourishment.

Species that normally give birth every two years rather than every year —langurs and baboons, for example—usually effect the weaning process before the first estrous period of the new cycle (Jay, 1963; DeVore, 1963).

TABLE 9.   Birth Periodicity in Primates

Macaques	1 year
Baboons	2 years (minimum of 14 months for chacma)
Langurs	2 years
Chimpanzees	1–4 years
Gorillas	3½ to 4½ years
Gibbons	2 years
Various lemurs	Probably yearly

Among chimpanzees and gorillas, the spacing of offspring ranges from two and one-half to four years (Goodall, 1965; Reynolds, 1965; Schaller, 1963, 1965), and among chimpanzees weaning probably occurs shortly before the birth of the new infant, although the actual cessation of nursing may be much earlier. Gorillas apparently do not nurse after about a year and a half (Tables 9 and 10).

Because of variations in primate birth periodicity and in the timing of the mother's rejection of her infant, the terms "infant" and "juvenile" should be applied with caution to the growing primates of different species. Among baboons and langurs, rejection at the first estrus shifts the infant into the juvenile stage relatively earlier than macaque rejection at the birth of the new infant. However, since these genera (macaques, langurs, and baboons) begin subadulthood at about the same absolute age (Table 10), their rates of growth and development are actually similar, and thus in those species with a two-year birth interval there is no need to extend infancy over a longer period of time.

Among howler monkeys, rejection of last year's infant is more gradual and usually does not occur until after the birth of the new infant. Since gorillas and chimpanzees have a wider spacing of infants, they are more relaxed about rejection, which is likely to be less traumatic for these infants than for baboon or macaque infants. Howlers, gorillas, and chimpanzees lack the strongly marked dominance system of the baboons and macaques and also have more relaxed attitudes toward their infants; whether there is a cause or effect relation between the two is debatable.

The two distinct stages of infancy, early and late, have probably evolved as much from group attitudes as from any inherent quality in the infant. The newborn infant is a curiosity and an attraction to the rest of the society, often marked by a contrast in color and behavior from the later infant, which is more likely to be treated as a nuisance. Although the older infant is still an object of adult grooming and protection, it is no longer tolerated as before. For example, a mother langur will carry her dead dark infant for days or even weeks, and a baboon male will insist that the female carry her dead dark infant through at least one day. On

TABLE 10.   Primate Life Stages After Infancy

Species	Juveniles	Subadults	Adults
Bonnet macaques	Young 1–2 years; older 2–3 years	♂ 3–7 years; puberty about 3½ years	♀ 3½–4½ years + ♂ 6–7 years +
Rhesus macaques	1–3 years	♂ 3–6 years; puberty 3½–4½ years	♀ about 4 years + ♂ 6 years +
Pig-tailed macaques		Puberty 50 months	
Toque macaques		Puberty 2½–3 years	
Savannah baboon	Young 12–24 months; older 2–5 years	Puberty about 4 years	♀ 4 years + ♂ 7–8 years +
Hanuman langur	15 months to 4 years	♂ 4–7 years; puberty 3½ years	♀ 3–4 years + ♂ 6–7 years +
Gibbon		Puberty 8–10 years	
Orangutan		Puberty 8–10 years	
Chimpanzee	Young 2–4 years; older 4–8 years	♂ 6+–10 years; puberty 8–9 years	♀ 6–8 years + ♂ 10 years +
Gorilla	3–6 years	♂ 6–10 years (blackback); puberty 7–10 years	♀ 6 years + ♂ 10 years + (silverback)
Man		Puberty 12–15 years	
Howler monkey	J–1: 20–30 months J–2: 30–40 months J–3: 40–50 months	♂ 4–6 or 8 years; puberty unknown	♀ 4–5 years + ♂ 6–8 years +
Marmoset		Puberty 14 months	
Bushbaby (galago)		Puberty 20 months	

the other hand, an older infant that dies or is killed may be examined by the adults but it is left behind.

The Juvenile Phase

Juveniles are characterized by a need to explore wider social relationships and the world around them while at the same time maintaining some ties with their mothers. The new juvenile, which is no longer the major interest in its mother's life, is displaced from close, regular contact with her. Thus it turns more and more to its peer group for social interaction, and the priority of mother and peer group reverses at this point. The juvenile, which already in late infancy had spent a great deal of time playing with age-mates and older juveniles, begins to spend even more time away from its mother. But a reluctance to sever the ties completely often prompts the juvenile to sleep with its mother at night and during resting periods in the daytime. Moreover, despite the ever-weakening ties between the mother and child and the deepening social relations with other members of the group, younger juveniles tend to remain in the same general section of the troop as their mothers.

The juvenile period of life is often clearly divided into two or three subdivisions. However, it should be noted that these subdivisions are a result of the birth season and do not indicate any marked transitions in the life of the juvenile primate. In those species that lack birth seasons because they live under conditions in which such seasons are unnecessary, it would make no particular sense to subdivide the juvenile stage. Thus among those monkeys and apes whose births are concentrated into a birth season and which take more than a year or so to reach adulthood, the juveniles fall into more or less clearly demarcated age classes of one year each. For example, among bonnet macaques, *juvenile one* is between one and two years of age, *juvenile two* is between two and three, and *juvenile three* is between three and four years of age, and at any point in time they are fairly easily distinguishable. The behavioral differences among these juveniles are part of a developing continuum rather than a sharp demarcation of behavior.

Juvenile one is still fairly closely associated with its mother, juveniles two and three are less and less so, particularly the males, which, as they approach adulthood, tend in some species to move to the edge of the troop and even to become isolated outside the troop. Many female primates remain in close association with their mothers throughout life, so the change in relationships between the female juvenile and her mother is often slight. Juvenile ones do not play as roughly as older juveniles and, when involved in a play group of mixed ages, tend to drop out as the playing gets rougher. The younger juvenile is more likely than the older to have the support of its mother if it is threatened by other monkeys, another indication of the slowly weakening bonds between mother and offspring.

Thus the basic characteristics of the juvenile period are increasing independence from the mother; increasing involvement in peer group relations; increasing socialization into the group, or, for some males, decreasing association with the group as the individual tends toward the periphery of social relations.

Subadulthood

Among most primates, the female, unlike the male, does not go through the period often called subadulthood but socially passes immediately from the juvenile period to full adulthood, that is, at the end of her third year for bonnet macaques or the end of her eighth year for chimpanzees. After puberty, she enters the reproductive cycle and with the birth of her first infant assumes the role of a socially mature adult female. According to Booth (1958), many females probably lack full adult dentition when their first infant is born, like many human females who give birth before

PLATE 9   Female Celebes ape (*Cynopithecus niger*), which contrasts strongly with the male (Plate 10) in size and robustness. (*Courtesy of the Oregon Regional Primate Research Center; taken by Harry Wohlsein.*)

their third molars have erupted. Thus the female primate tends to achieve social maturity before complete physical maturity. Her first menstrual period is a good indicator of imminent adulthood.

The male primate, on the other hand, becomes physically mature long before he becomes socially mature. His juvenile period ends with the eruption of the full set of permanent teeth and the onset of sexual maturity, though these do not precisely coincide. The obvious, easily observable dental sign, the eruption of the large, projecting canines, occurs

PLATE 10   Male Celebes ape. Note the markedly heavier face and shoulders in contrast to the female (Plate 9). (*Courtesy of the Oregon Regional Primate Research Center; taken by Harry Wohlsein.*)

while the second and third molars are still erupting. At this point the male and female are usually about the same size, but the female of many species soon ceases to grow, whereas the male continues to grow until, in the case of baboons, macaques, gorillas, and some other species, he is twice the size of the female. However, not all males reach full size: often a significant number are little larger than the females. Among arboreal species, sexual dimorphism tends to be minimized. For example, size differences and other anatomical secondary sex differences are almost totally lacking among gibbons (Carpenter, 1940). However, the respectable sexual dimorphism in size among howlers, which are also arboreal, weakens the generalization. Chimpanzees, which are as ground-dwelling as many macaques, have only a slight sexual dimorphism in size, the female range overlapping that of the male. The attempt to link dominance in a species with strong sexual dimorphism (for example, among the baboons) likewise fails since the mountain gorilla, despite his weak dominance system, displays strongly marked sexual dimorphism (Schaller, 1963). Thus the most that can be said without further data is that the causes of sexual dimorphism are complex.

During the period of subadult growth, which usually lasts three to four years, the male is sexually mature and capable of reproducing but is not socially mature and, in some species, is usually excluded from access to the females. In species with a strong dominance hierarchy, the male while gaining in size and strength begins to move up in the hierarchy or leaves the troop and becomes isolated. Later, if he returns and attempts to work into the dominance hierarchy at a high level, he may or may not succeed; at any rate, he is more likely to have been picked off by predators when he was isolated from the troop. If he survives, his other alternative is to move to another troop and thus aid gene flow from group to group. Among species where isolation of subadult males is the rule and where predation is significant, the death rate of males must be higher than that of the females. The sex ratios of many baboon and macaque troops seem to support this assumption, the ground-dwelling species tending to have a high percentage of females.

The social interactions of the subadult male are more serious, and he plays less than the juvenile. When he does play, he is usually rougher and comes closer to actual fighting. As older animals play more roughly, younger ones drop out of the play and watch. This rowdy play probably merges into a taunting of subordinate adult males by the subadults that are beginning to work their way up the dominance hierarchy.

Because of the age disparity when males and females reach full adulthood, in the average primate troop the adult sex ratio usually favors the females.

The Adult Male

The male is socially adult when he has attained not only the size, strength, and secondary characteristics of physical and sexual maturity but also a recognized position in the social organization of the troop. Thus, a physically adult isolated male has not reached full adulthood because he has not completed his social maturation. A fully adult male is an active social arbiter in the routine of the troop. He has a role to play in the general leadership activity of the troop: protecting the troop, assisting females to ward off the undesired attentions of subadult males in some species, help-ing to maintain the order of the group. The most dominant males, of course, assume the most responsibility for group leadership, but the lower-ranking males, if the group is large enough to have them, are usually assigned some kind of role. A more complete discussion of adult roles is found in Chapter 7.

Old Age

Unlike some well-known human societies, subhuman primate societies put no premium on youth for youth's sake. An old female is as likely as a young female to be supported by a young dominant male, often more so because she has probably established a high social status not yet attained by the younger female. Moreover, an old female is less handicapped socially by old age than the male since she does not rely on her own re-sources for defense and support but often calls on the dominant male for help. In some respects, the female relies more than the male on wile to gain and retain her position, and her wile is likely to increase rather than decrease with age! Very old females often give birth regularly; only the human female among primates appears to experience the menopause. In fact, there is no indication that age per se precludes participation in the reproductive cycle for either sex. However, old age generally means a lessening of the powers of the individual. Canines, for example, tend to be lost with other teeth, often earlier than any of the others. In species that rely heavily on dominance, the old male tends to drop somewhat in influence but by no means universally. Although a highly dominant male often becomes less so in old age, he probably never loses all of his influence.

There is, of course, a steady attrition at all stages in the life cycle. The heaviest death rate occurs during the earliest part of life. Probably more young juveniles, those entering the first period of full rejection by the mother, are lost than those in any other age group. Exposed to the weather and predators but lacking the care given by the mother during infancy, they are not yet old enough to have built up sufficient resistance and strength to meet emergencies. Nature's selection for resistant animals is strong at this time.

Subadult males represent another group characterized by high mor-tality. Much like adolescents in our own society, they indulge in rash behavior that frequently endangers their lives and do not take the pre-cautions that older, and presumably wiser, monkeys and apes do.

A good example of such rash behavior is seen among the bonnet macaques of southern Mysore State, India, which live along the banyan-lined roadways. Subadult males often sit along the edge of the road and, hearing a car or truck approaching in the distance, turn away from it, ignoring it until the last moment when they jump aside or merely casually swing their tails out of the way. Sometimes such casualness doesn't pay off! No adult macaques or even young females indulge in such behavior.

TABLE 11.   Some Primate Longevity Records

Golden-headed saki	14+ years
Hooded capuchin	33+ years
Kikuyu colobus	20+ years
Proboscis monkey	4 years
Galago crassicaudatus	12 years 8 months+
Nycticebus coucang	8 years 2 months+
Chimpanzee	24+ years; 39 years

It is the casual flippancy of youth compared with the wiser awareness of adulthood. Subadult males appear to seek out danger, whereas adults expose themselves to danger only when maintaining troop protection or increasing their status makes it necessary to do so. When subadults do take part in troop protection, they often take unnecessary chances and thus jeopardize their lives.

It is exceedingly difficult to determine the age of adult primates in the wild. Only through observations over a long period of years can the average age of a species in any given area be reliably known. But in a laboratory situation, it is comparatively easy to determine the age of subadults and younger animals since these categories can be worked out over a few years and applied to the wild animals. Obviously, however, longevity in the laboratory is based on different factors from those in the wild. Once established in a laboratory colony, disease is likely to sweep through the whole population; predation, on the other hand, is not a factor in the laboratory. Since the free-ranging primate group comes into direct contact with other groups only rarely, it is better able to avoid disease, but it is, of course, subject to local predation.

We can only estimate the longevity of individuals in the wild; Tables 11 and 12 give some authenticated ages for various animals under controlled conditions. Many adults probably die in what would be equivalent to young adulthood in man, that is, the late teens and twenties. Many others succumb before middle age; only the highly exceptional animal reaches true old age in the wild. The mortality pattern found among human hunters and gatherers is probably analogous to that of the subhuman primate in the wild.

For macaques, fifteen to twenty years probably represents an average maximum life expectancy. That some live past thirty even in the wild is shown by Jupiter, the mature troop leader in the Takasakiyama troop of Japanese macaques. When observations began in 1948, Jupiter was estimated to be fifteen years old. He was well established in a stable dominance hierarchy that included several classes of males; he and six others formed the dominant class. If the 1948 estimate was correct, then he was

TABLE 12.   Selected Longevity Records

Shrew	2 years
White rat	4+ years
Earthworm	5–10 years
Rabbit	12 years
Sheep	15+ years
Ant queen	19 years
Large spider	20 years
Dog	24+ years
Domestic cat	27+ years
Lion	30–35 years
Pelican	40–55+ years
Horse	40+(?) years
Hippopotamus	49 years
Termite primaries	40–60 years
Indian elephant	77+ years
Golden eagle	80(?) years
Sturgeon	80–100+ years
Man	115+ years, 168 years
Vulture	117+ years
Large tortoises	100–150+ years

Taken primarily from Comfort, 1964.

nearly thirty when he died in the winter of 1960. Titan, the second-ranking male, which could well have been fifteen or thereabouts in 1948, survived Jupiter by several years. An old male in the Somanathapur troop of bonnet macaques in Mysore State, which showed signs of very great age far beyond that of other obviously old males, may well have been nearly thirty years of age, if not more. These are all estimates, however; only the kind of continuous observations that have been carried out at Takasakiyama since 1948 will provide accurate data on longevity.

Despite these examples of old age, many circumstances in the wild—disease, predation, falls, and fights—make it difficult for the primate to achieve old age. The potential for a long life apparently is present, but because of the exigencies of life in the forests and savannahs it is seldom attained.

THE CYCLE OF DAILY ACTIVITY

The daily cycles of primate activity fall into two distinct patterns, diurnal and nocturnal. Most living prosimians are nocturnal; most higher primates diurnal. Among the former, the only exceptions today are found on Madagascar, where they have no competition from the more successful monkeys, or among the tree shrews, whose way of life is quite different from that of most other primates. The galagos, pottos, and angwantibos of Africa, the lorises and tarsiers of southern and eastern Asia are all nocturnal and live in the same general area in which monkeys and apes are found. Only the lemurs of Madagascar include both nocturnal and diurnal species (Petter, 1965). "In the Lemuridae, among the subfamily Cheirogaleinae the genera *Microcebus, Cheirogaleus,* and *Phaner* are strictly nocturnal. After nightfall these animals emerge from cover, which may be a hole in a tree or a nest in foliage, where they have spent the day, and they do not return until the first light of day." In their active periods, the genera *Lemur* and *Hapalemur* are intermediate between nocturnal and diurnal forms. They are most active during the periods of half-light, the early morning and evening hours between full daylight and full darkness. They usually spend the night sleeping and the day resting, but they do move about during the day on occasion and are capable of full activity at night if disturbed. The third genus of the subfamily Lemurinae, *Lepilemur,* is fully nocturnal and leaves its cover after dark, returning before daybreak. The genus *Avahi* of the family Indridae and *Daubentonia* are also strictly nocturnal.

The patterns of daily activity of the nocturnal forms are not well known simply because it is difficult for the diurnal human species to observe a nocturnal form. Hence most of the following discussion will center around the diurnal species. Let the reader remember, however, that this emphasis is due first to the lack of observations of comparable intensity of living nocturnal forms, and second, to the fact that the majority of living primate forms are diurnal.

THE DIURNAL PATTERN

The coming of dawn signals the beginning of activity for diurnal primates. Sleeping groups break up as the monekys and apes begin moving from one position to another in the trees or sleeping areas. Infants are nursed and some mild play often begins. The species spacing calls and signals are made at dawn with the first burst of activity. The gibbons of the Old World and the howlers of the New are unquestionably the most vocal users of spacing calls. Carpenter (1940) noted that "These [gibbon] calls stimulate groups in adjacent territories which give a series of similar notes and the contagion of sound spreads from one group to another until each has called repeatedly during the next hour." The impressive roar of the howler monkey is most common at dawn, and those of the gibbon and howler seem to be a means of establishing the relative locations before they move to the food trees. Whereas this behavior is most spectacular among the above-mentioned species, signals given by many other primates may well serve the same purpose. The coughing whoop of the male Nilagiri langur, often heard in the early morning hours, and the branch-shaking of the macaques are used to locate different groups or separate parts of a single group. For example, a troop of bonnet macaques was temporarily divided and spent a night in two sleeping trees three-quarters of a mile apart. At dawn, a large male in one section shook the branches of a tall tree long and hard; a little later, the other part of the troop arrived and joined the others as they moved into the food trees.

Although most primates begin their activity with the first light of dawn, some begin to move about even before it is light. Nissen (1931) reported chimpanzee activity before daylight, and bonnet macaques have been observed before daybreak moving from a sleeping tree to an eating tree by the light of a full moon. However, activity usually coincides with the first light, though in some circumstances the animals remain in their sleeping positions or in that general area until full daylight. Baboons, which spend most of their waking hours on the ground in areas with large predators, do not generally leave their sleeping places until the dusky shadows of morning are dissipated, usually about an hour after dawn (Hall, 1960, 1962a). Gorillas often remain in their nests for half an hour to an hour before beginning to move away, led by the dominant male (Schaller, 1963). Probably because of the generally low level of predator pressure in the trees, arboreal species usually begin their activity with the first light. Lumsden (1951) reported that *Colobus* and *Cercocebus* monkeys leave their sleeping places during the period before full sunrise. Bonnet macaques (Simonds, 1965) move out of their sleeping trees before sunrise but do not reach the ground until full daylight. The common Indian langur generally follows the same pattern (Jay, 1963) as do howlers, gibbons (Carpenter, 1934, 1941), and other macaques.

The time of first activity often varies with temperature and light. Carpenter (1941) stated:

During foggy, dark and cool (45° F) mornings on Doi Intanon, the morning calls were not heard until about seven o'clock and the groups started moving about eight o'clock. On clear mornings with temperatures around 65° F. at Doi Dao, the animals would start their morning calls near five-thirty and begin progression between six and seven-thirty o'clock.

After a rainy night, bonnet macaques are less active and slower to move from the sleeping trees even if the sun rises in a clear sky. Schaller

(1963) reports that clear or cloudy days did not affect the time when gorillas first began to move away from the nest. Perhaps those primates that sleep high in the trees, where they are easily affected by the light, respond more readily than those which, like gorillas, sleep close to the forest floor.

The Morning Feeding Period

Depending partly on the weather and partly on the availability of food, the movement from sleeping to feeding places varies. On cool mornings, feeding often lasts for a considerable time and is interspersed with other activities, so that the rest period does not begin until late in the morning. When the weather is hot and sultry, feeding begins early, and if there is plentiful food, terminates early so that the rest period of the later morning and early afternoon fills the long, hot hours of midday. If the animals must move over wide areas to collect enough food, the feeding period is longer.

Gorillas (Schaller, 1963) normally need about two hours of intensive feeding to satisfy their requirements before resting. They generally feed shortly after they leave the nest. Macaques and langurs also feed in the early morning, spending the late morning resting. Gibbons (Carpenter, 1941) spend the early morning moving about and usually do not settle down to feeding until the late morning. New World howler monkeys show a similar pattern of progression before feeding. In contrast, I have seen bonnet macaques move quickly into a tree at five o'clock in the morning and begin active feeding. Although this is an exception, their progression rarely lasts the hour or two that Carpenter reports for howler monkeys and gibbons.

Another factor, the location of food in relation to the sleeping places, determines how much travel is necessary between rising and settling to the morning meal. During different periods of the year, the pattern for a single group tends to change, particularly if they return to the same sleeping places night after night. The morning feeding period, coming after a nightlong fast, is an intense one in which the whole group is involved. For the first part of the feeding period, other activity—play, chasing, sexual activity—is usually at a minimum. But as the animals become satiated, they slow down in their eating and begin to engage in other behaviors. Infants and juveniles begin to play, and adults settle down for short bouts of resting and grooming. Consort pairs of males and females move away from the center of the troop to mate and groom; in many species they sometimes mate close to the other members.

When more than one species of primate occupies an area, two species are often intermixed in a single tree or other feeding area, generally peacefully and without conflict. In many forests of the Old World, deer and other ungulates often feed below the trees in which monkeys are plucking fruit or leaves. Since the monkeys drop a good deal of uneaten food, the browsing is quite good for the ungulates. On the savannah, baboons and impala or other ungulates often move and feed together, usually for mutual protection: Since the baboons have better sight and the ungulates a better sense of smell, it is almost impossible for a predator to surprise the combined group.

Midday Resting

The midday resting period follows feeding. As the temperature rises, the primates move into cool, shaded resting places and remain there during the hot period. In hot lowlands, they spend the whole midday period in one area, scarcely moving about. In the cooler higher elevations, the midday activity is less muted and there is more likelihood of movement from place to place and participation in play, grooming, and feeding. Hot, sultry weather acts as a damper on all activity; in fact, temperature often determines the position of the animals in the trees or in their home range generally. In the cool of the early morning, monkeys tend to move to the sunny side of the trees and remain where the warmth of the sun's rays can strike them. As the heat increases, they pull back and find resting places in the cool, shady recesses of the trees.

Except for gorillas, which seem to rest near or in the feeding area, the animals rarely rest in the same trees or areas in which they have just fed; instead, they move to favorite places that they regularly use for midday resting. These places are generally well protected or so situated that the animals can notice predators at a distance and take the necessary measures to protect themselves. Gorillas, which have few natural predators, spend most of their time on the ground.

During the resting period, the major activities are dozing and grooming for the older monkeys and apes and play for the infants and juveniles. In some species, particularly baboons and macaques, the troop often separates into small clusters of individuals whose membership is generally stable over periods of months and years. The same females that congregate and groom one another probably represent mother–daughter relationships (see Chapter 7) or special friendships. In other species such as gorillas and chimpanzees, the preferential arrangement seems to be lacking, except perhaps for the silver-backed male in a gorilla group.

Partly because the play activity of the young of many monkey species occurs in the same places near the resting area, it often seems stereotyped. They jump, leap, and climb in the same bushes or branches day after day.

Midday resting is one of the most relaxed periods of the day; usually there is little or no tension within the group. For the observer, particularly of arboreal species, it can be an exceedingly boring period: many of the animals are hidden from view by the foliage and are, for the most part, doing very little anyway.

Afternoon and Evening Period of Activity

The long midday siesta sometimes extends into the afternoon or is broken into several resting periods interspersed with short periods of activity. *Lemur catta* (A. Jolly, 1966), like many other species of primates, breaks

the siesta with feeding periods. The length of their siesta varies in relation to the climate. Jolly states that *L. catta* sleeps from about noon to
4 P.M. in the hot season and for a shorter time in the cold season. Rainfall
during the day or night also affects the length of the siesta. Bonnet
macaques are less active and tend to rest longer in the midday after a
night of heavy rains and during heavy daytime rains. Gorillas, too, are
affected by the weather and sleep longer on warm, sunny days than on
cool, cloudy days (Schaller, 1963). Thus when the late afternoon period
of feeding and activity begins depends on several factors and tends to
differ from day to day and season to season.

Typically, in the late afternoon primates move from their resting places
to a feeding area where the whole group feeds intensely for various
lengths of time. Gorillas require so much food that their feeding periods
are longer and more intense than those of many other species (Schaller,
1963). Baboons that live where food is sparse or widely dispersed have
shorter resting periods and spend longer hours feeding and moving in
search of food than those living where food is more plentiful (Hall,
1961). The rhesus macaques that feed on sheesham seeds must spend
virtually all waking hours during the season picking the seeds from the
complex pods. But species and groups within species that live in areas
of abundant food spend as little as two or three hours in intensive
feeding.

A few species (gorillas and red-tailed monkeys—Schaller, 1963; Haddow, 1952) spend more time feeding in the afternoon than in the morning, but the afternoon feeding tends to be less intense and continues
sporadically for the rest of the day, whereas the morning feeding is intense and concentrated. It terminates for bonnet macaques, for example,
as the troop as a unit shifts to the resting trees. By contrast, their afternoon is a mosaic of feeding and other activities. For the troop as a whole
one cannot say when feeding leaves off and other activities begin. The
bonnet macaques move slowly toward the sleeping trees, eating, playing
and interacting socially. Feeding itself sometimes causes conflicts among
the individuals. As the troop moves slowly, infants and juveniles play,
foraging intermittently. Subadult males in large groups sometimes bait
older, low-ranking adult males. Sexual activity increases. A large troop
(fifty or more) becomes markedly more active. Bonnet macaques move
about, play, and chase along the ground. *L. catta* and *Propithecus verreauxi* (A. Jolly, 1966) take longer to progress from eating to sleeping
areas than in the morning. Baboons intersperse their eating with play
and grooming before moving into the sleeping trees. Perhaps because
their groups are smaller than those of the terrestrial macaques and baboons, gorillas are less active than many other primates. Once the play
and threat behavior is started, it is not intensified and reinforced by the
addition of other gorillas. Large troops of baboons and macaques, however, often give the impression of increased and varied activity during
the later afternoon and early evening.

Sleeping Activity

As the monkeys and apes move into their sleeping areas, their activity decreases markedly but by no means ceases. Ground-dwelling and semi-ground-dwelling monkeys move into sleeping trees before sunset or shortly thereafter, but once in the trees there is considerable shifting among the adults and play among the younger animals before they take their sleeping positions. Although predator pressure makes it dangerous to remain on the open ground, their activity can continue in the trees.

It is probably quite common for troops of arboreal monkeys to sleep in widely separated trees at night, but it is rather rare for terrestrial species to do so. Unless the troop undergoes division, a baboon troop chooses a single sleeping area each night, and the somewhat more arboreal bonnet macaque troops rarely sleep in two or more distantly separated sleeping trees during the same night. Lumsden (1951), however, reported that twelve times in a total of thirty-five observations, the lowland colobus monkeys slept in two or more trees, on four occasions the distance between sleeping trees being 40 to 120 meters.

In areas occupied by several species of monkeys, some differences in times of settling for the night have been noted. In the Semliki forest, the colobus monkeys move into the sleeping trees before either the mangabeys or baboons but do not take up their actual sleeping positions until sunset. At first they move about in the lower parts of the crown and then move into the high crowns of the trees for the night. Here, also, species preferences for sleeping positions vary. The black mangabey sleeps among the small branches in the peripheral crown, whereas the lowland colobus prefers the small branches of the main crown. Redtails usually sleep among the small twigs; the baboons prefer large, nearly horizontal branches close to the tree forks and therefore lower than the spots selected by the other three species. If there is heavy rain at sunset, baboons in the Semliki forest sleep on the ground (Lumsden, 1951).

Despite their apparent safety, the trees and cliffs used for sleeping at night are open to some predators, and watchfulness is still necessary. Disturbance by predators must be great to cause the animals to shift their sleeping places at night. Hall (1962a) and Zuckerman (1932) indicate that only rarely does the approach of human beings cause a troop of baboons to leave its sleeping cliffs. The sleeping place was originally chosen for its security from predation; were the animals to move out at the approach of the predator, they would probably only increase their danger.

Many Old World species of primates have evolved ischial callosities as an adaptation for resting in the sitting position. These enable the monkeys to sleep safely on small branches away from the trunk where their tree-climbing predators cannot easily reach them. In addition, the smaller branches are easily disturbed by the weight of an approaching predator and thus serve as an alarm to the sleeping monkeys. Baboons that sleep on rocky cliffs have, of course, lost this alarm feature, but it is otherwise an efficient night signal for diurnal animals that lack effective night vision. Apparently some diurnal lemurs (*L. catta*, A. Jolly, 1966), unlike diurnal monkeys, have a relatively efficient nocturnal vision since they have been seen to leap from branch to branch even on dark nights. The alarm feature of sleeping on small branches still serves to awaken a sleeping primate.

The choice of the sleeping place depends upon the local availability of suitable trees or substitutes for them. By far the majority of species are exclusively arboreal and their sleeping places are confined to the trees themselves. For baboons in particular, however, the nature of the sleeping place varies considerably. Hall (1962a) reports having seen them resort to vertical cliffs rising out of the sea near the Cape of Good Hope. DeVore (1965) describes them using fever trees as night resting places in Kenya and Tanganyika. Kummer and Kurt (1963) report that large aggregations of hamadryas baboons sleep on cliffs. The similar, but only distantly related, *Theropithecus gelada* also takes to the precipices of the Ethiopian highlands for sleeping places (Crook, 1967). In each of these cases, the sleeping place must be climbed to be reached and hence is generally inaccessible to predators.

Many macaques, having made another kind of adaptation, move into towns and temples where as often as not they resort to inaccessible niches in the buildings of the town to sleep. Only the gorillas habitually sleep on the ground, and few are the predators capable of taking on a group of gorillas!

Since animals over 100 pounds find it almost impossible to secure safe arboreal sleeping places, the great apes have devised their own specialization. Chimpanzees, gorillas, and orangutans all construct sleeping nests, that is, a platform of branches that have been bent back on themselves to form a crude, interlacing network of small and medium-size branches. Here the animal can lie without much danger of falling.

The highly arboreal orangutan sleeps nightly in nests built far above the ground. Although chimpanzees spend much time on the ground as well as in the trees, most of their reported nests are in the trees; Reynolds (1965) found only two ground nests. Since gorillas are basically terrestrial today, they construct 90 percent of their nests within 10 feet of the ground or on the ground (Schaller, 1963). For the gorilla this nest construction seems to be a retention of behavior patterns common to ancestral apes.

Virtually all other higher primates and many of the prosimians sleep sitting up in the trees, often huddled together in groups of two to ten animals. The group usually sleeps in the same general area but many times sleeps in adjacent trees and is thus subdivided into small groups. Since nighttime observations in the forest are rarely rewarding, accurate data on which individual sleeps with which are rare. However, on several different nights a small group of bonnet macaques was observed sleeping in the same group of bamboo clumps. The single male of the troop, which was easily recognized, slept consistently on the same bamboo twig each night. On one night, an older infant male slept huddled with him and on another night an estrous female shared his twig.

Although the nocturnal primates are not known sufficiently well to be adequately described, their sleeping places have been recognized in some cases. Except for the inhabitants of the island of Nosy-Be, the genus *Lepilemur* spends the day in holes in trees. On Nosy-Be there are no mammalian predators to pose a threat, so they simply roll into a ball in the middle of the foliage (Petter, 1965). The nocturnal *Avahi* also sleep like the Nosy-Be *Lepilemur*.

Shifting during the night is common among non-nesting primates, most species reportedly moving about in the branches of the sleeping tree. This seems to be the result of mild altercations among themselves, recognizable by growls and screeches or equivalent vocalizations.

Whether the animals spend their nights in the same sleeping spots or not seems to be a matter of group tradition. Primates living in areas where sleeping places are limited generally return to the same or one of a few sleeping places night after night, since there is little choice. Nest-building primates, on the other hand, tend to build a new one every night and reuse of an old nest is rare. Variation within the species has been noted. Nolte (1955) reported a troop of forest bonnet macaques sleeping in different places each night, whereas I observed a smaller forest troop of the same species returning regularly to the same stalks of bamboo, at least during the monsoon season.

ANNUAL AND MULTIANNUAL CYCLES

Seasonal changes and cycles of vegetation changes that exceed one year affect the behavior of primate populations. The alteration of food sources within the group's home range makes it necessary for the troop to shift its daily pattern of movement to make use of the newly available food sources. For example, gorillas make greater use of bamboo as a food source during the periods of heavy rain, and although it is not a major shift, it does cause an alteration of the daily pattern (Schaller, 1963).

Dry seasons often reduce the food sources of a population to a marginal level, sending them farther from their favored areas and increasing the danger of predation. In some cases, major shifts in home range take place seasonally. During the dry season, a troop of *Macaca radiata* in southern Mysore State shifted entirely out of its monsoon and cool season home range and returned after the rains had started again. In some cases, intergroup contact is increased, as Jay (1965) reports for the Hanuman langur. During the dry season, two or more groups that would otherwise rarely come into contact meet at water sources. Similar behavior is reported of baboons in the Amboseli Reserve in East Africa, where during the dry season they drink at the few remaining water holes; several troops come together at such times but retain their troop unity.

Hot weather reduces the activity of bonnet macaques, which seek the cooler recesses of the trees rather than continuous activity in the hot sun. Gorillas are seemingly unaffected by the heat, and they continue their usual pattern of behavior when it is hot. However, if it is both hot and dry, primates may well be required to expend extra energy in gathering food. The food requirements generally take precedence over extremes of temperature.

Rainy seasons too tend to curtail the daily activity cycle of primates. In periods of heavy rain, the monkeys huddle in protected places and resume activity only when the rain slacks off. Lowered temperatures can cause periods of torpor similar to hibernation in some prosimians—*Cheirogaleus major, C. medius, Microcebus murinus* (Petter, 1965).

In general, however, the alternations caused by annual changes in temperature are minimal, and the primate pattern of daily life continues in the same vein, with minor adjustments to changing conditions. The major effect of marked seasonal variations in rainfall is seen in birth and mating seasons. If there are lean seasons caused by drought in the annual cycle, infants are generally born in a restricted season of two or three months and become dependent upon extramammary food sources when the rains have produced fresh growth in the vegetation.

Cycles longer than a single year also affect the even tenor of primate life. In southern India, the bonnet macaques that have not been forced to adapt to man's alterations of the environment live primarily in bamboo forests. Bamboo has a life cycle of about fifteen years; it produces edible shoots during the monsoon season and provides havens from predators all year. At the end of the fifteen-year cycle, the bamboo goes to seed and dies—sometimes in vast tracts—and it is several years before the region can again support the maximum population of bonnet macaques, which are then forced to move; hence major readjustments of the local population must be made every fifteen years. In Gundlupet Taluk of Mysore State, the population of roadside monkeys is augmented every fifteen years as the forest monkeys shift northward, seeking alternatives to their depleted tracts of bamboo.

Cycles of epidemic diseases also cause readjustments in the local primate populations. Studies made on Barro Colorado, Panama Canal Zone, show that yellow fever takes a periodic toll of South American monkeys. There during the late 1930s the population was reduced drastically but climbed again in the 1950s.

Man's activity also produces cyclical changes in the environment. In parts of Africa, regular burning of the savannah forces the baboon and patas monkey populations to readjust in their search for food and their avoidance of predation.

Although most of these cyclical changes do not cause major problems for most primates, the periodic adjustments they entail tend to expose the members of the group to increased predator pressure and to disrupt the internal social organization, making it necessary for the animals to learn new patterns of behavior.

Chapter 5
Adult-Young Relationships

Much has been written about the close, intense, and durable relationship that develops between the primate mother and her infant. It is an exceedingly important relationship in primates as in many other mammals because of the relative helplessness of the newborn mammal, which requires a period of protected learning if it is to survive and function as an adult.

Although the mother–infant relationship is the main topic of this chapter, other relationships important to the developing primate will be treated here. In many ways, the relationship between adult males and the young is equally significant. Characteristically, in many primate societies, the males share a great deal of the responsibility of protecting the young and influencing their development; they are as much a constant in the young primate's environment as the females. Present in the group throughout the year, they play a leading role in the society. They are usually the social arbiters, the agents of social control. The complete socialization of the young primate depends, in part, on the presence of adult males in the social group.

No treatment of dyadic relationships alone, even of the male–infant and mother–infant dyads, gives a complete picture of the social and physical environment in which a young primate develops. The collections of individuals that compose a social group are an integral part of the social interactions that in primate societies are rarely characterized by simple dyadic relationships. At any given moment, an individual primate may be chiefly concerned with a particular fellow primate, but he is rarely unaware of the other members of the group. The infant's social awareness develops within the warm climate of complex social inter- action in a group whose members represent both sexes, all ages, and a rich variety of personalities. Thus a discussion of adult–offspring rela- tionships leads inevitably to a discussion of many other social relationships that profoundly affect the growing primate. These relationships are treated elsewhere in this book.

THE MOTHER–INFANT DYAD

The reciprocal mother–infant relationship has been studied in two quite different contexts. The laboratory studies of Harlow, Mason, and others have concentrated mainly on the affectional systems within the dyad. They have examined the different stages of the relationship and the pri- mary elements that help to maintain the relationship itself. Field studies, on the other hand, have concentrated more on the significance of the relationship for survival, its adaptive value, and the part it plays in the total social setting. Thus, whereas the laboratory studies have concen- trated on the dyad in isolation or within strictly controlled social limits, the field studies have focused on the setting, social and environmental, in which the dyad exists in a viable society. These two approaches are complementary; each sheds light on the mother–infant dyad that could not be generated by the other. Field studies cannot isolate factors that influence the dyad and test the strength of each; but neither can the laboratory situation duplicate the complex social and physical environ- ment in which the dyad evolved to its present condition.

Of all mammalian social relationships, only one—the sexual relationship necessary for conception—is more basic to the survival of the species than that between a mother and her infant. Most mammalian species could not survive without the kind of close relationship between the young and the mother that supports and sustains the infant in its most defenseless period of life. Some forms of animals do not have special social relationships between mother and infant; for example, several reptilian and amphibian forms lay their eggs and then desert them. Most of the newly hatched offspring, left to fend for themselves, do not sur- vive. Thus these forms must produce numerous offspring, only a few of which will survive the rigors of early life and reproduce in their turn.

By emphasizing and maintaining the maternal–infant relationship whereby the infant for a considerable period of its early life depends for survival on the mother, mammals have reduced the number of off-

TABLE 13.   Reproduction in Selected Mammalian Species

Species	Menarche or Puberty	Age of ♀ at First Parturition	Average Offspring per Pregnancy	Birth Spacing
Chimpanzee	8 yrs 11 mos (mean)	10± yrs	1	1–1.5 yrs
Chacma baboon	2.5–3.6 yrs	4.5± yrs	1	1.5–2 yrs
Bonnet macaque	2.5–3 yrs	4± yrs	1	1 yr
Lemur catta	2.5 yrs		1	1 yr
Mouse lemur			2–3	
Jackrabbit (*Lepus californicus*)	1 yr	1+ yrs	2–3	2+ mos
Grey squirrel (*Sciurus carolinensis*)	1 yr	1+ yrs	3–4	6 mos
Wood mouse (*Peromyscus leucopus*)	7–8 wks	10–11 yrs	4	1+ mos
Elephant shrew (*Elephantulus myurus*)	5 wks	13 wks	1–2	2+ mos
Domestic cat (*Felis catus*)	7–12 mos	12+ mos	4	4–6 mos
Domestic goat (*Capra hircus*)	1 yr	18 mos	2	6 mos–1 yr
Domestic horse (*Equus caballus*)	1.5–2 yrs	3 yrs	1	1 yr
Indian elephant (*Elephas maximus*)	5–13 yrs	7+ yrs	1	4+ yrs (?)

Taken primarily from Asdell, 1964.

spring. For example, primates have, for the most part, reduced the number of offspring to one per pregnancy. The typically wide spacing between infants also emphasizes that the survival of the species depends on the few infants that are born each year (Table 13).

With the evolution of higher forms, several stages of the mother–infant dyad have become increasingly complex and reliable. One of these is the physiological system that produces the new individual. Implantation of the fertilized egg and the development of the new individual in the uterus of mammals are definite advances over the typical reptilian system of laying the eggs in a harsh and hostile environment and then allowing them to hatch untended or, in some species, guarded by a parent. Moreover, the system of nourishing the embryo became more complex as forms relied on fewer and fewer offspring to perpetuate the species. Among lower, more primitive primates that have a bicornuate or two-horned uterus, which usually produces two offspring per pregnancy, the placenta is epitheliochorial, that is, it retains several layers of tissue in both the uterine wall and the placenta itself, through which oxygen and food must pass to the developing individual and through which wastes must pass from it to the mother. Human beings, on the other hand, as representatives of the advanced higher primates, have hemochorial placentation, in which the whole wall of the uterus is eroded so that the blood of the mother passes directly over the placental membrane and thus facilitates the exchange of oxygen, food, and waste materials.

Twinning is relatively rare among monkeys, apes, and human beings, which have the more efficient systems of nourishing the embryo and fetus. Each higher primate is a highly complex organism that requires a long period of development to reach maturity. It would be exceedingly wasteful of the species resources if most of such offspring were to die; the higher primates concentrate on producing one at a time efficiently.

Just as the physiological relationship is more complex among those forms that produce fewer offspring, so is the social relationship between mother and infant. Laboratory studies by Harlow, Mason, and others have shown that this relationship not only has survival value for the species but also is necessary for the proper development of the newborn primate into a normal social being. Since the chief mechanism for survival among primates is the social group, the mother–infant relationship assumes a wider importance because it lays the foundation for the infant's introduction into a social milieu. Although such orientation may be foreshadowed in other mammals, it is not developed to a high degree. A primate raised in isolation from other primates, from its mother and other young and adults of both sexes, is not a primate in the social sense. Even a primate raised with its mother alone is sadly lacking in social finesse. Since the mother–infant relationship is one of the building blocks of the primate society, it is important to understand as much of the interaction as possible. Later chapters will show how the relationship can be modified to produce different forms of primate societies. Currently only a few species have been tested in the laboratory in such a way as to show the development of the mother–infant bond step by step. The study discussed below is based on work with the rhesus macaque (*Macaca mulatta*) with occasional references to other species. In considering the development of this interrelationship, note the strength of the bond that is formed between mother and infant. It is this bond that is important as a building block for the larger and more complex societies that develop.

MOTHER–INFANT SOCIAL RELATIONSHIPS

In a series of experiments on rhesus macaques, Harlow, Harlow, and Hansen (1963) concluded that two separate affectional systems are involved in the mother–infant dyad: the affectional system of the infant for the mother and that of the mother for the infant. The development of each system is independent of the other; for example, a monkey mother can lavish attention upon a dead infant, which can neither respond to her attentions nor give cues to sustain the relationship—but only for a few days or weeks at the most. On the other hand, an infant tries again and again to make contact with a mother that has rejected it. The full development of their relationship, however, depends upon reciprocity; both the mother and the infant must go through the stages together, respond to each other, and reinforce each other's patterns of interaction.

THE MOTHER'S RESPONSE TO HER OFFSPRING

The experiments conducted by Harlow *et al.* involved the use of what they call the playpen situation. Four separate living cages were placed adjacent to a play area that could itself be divided into four separate areas. Each living cage had an opening into the play area that allowed the infant macaque but not its mother to pass through. Within twenty-four hours after birth, mothers and their infants were introduced into this situation and remained there until the infant's ninth month.

Each infant was allowed two one-hour periods daily of direct contact with one of the other three infants five days a week from the 16th to the 180th day. The partitions were lifted and the two infants could interact with each other and with either of the two mothers. The positions were shifted on a predetermined schedule so that each infant interacted with all of the other seven individuals in the experimental situation.

Stage I—Maternal Attachment and Protection

As a result of these experiments, Harlow *et al.* described three stages in the development of the rhesus macaque's affectional system for her infant. The first stage is that of maternal attachment and protection in which "the mother monkey spends a great deal of time holding the baby close to the ventral surface of her body or cradling it loosely in her arms and legs, but still providing it with active contactual support." In addition, the mother spends much of her time grooming and restraining the infant from leaving her or retrieving the infant when it succeeds in moving a short distance away. The mother also directs numerous threats at the experimenter and thus indicates a strong protective response. Nursing is not mentioned as a factor in this period because it is as much a matter of infant initiation as it is of motherly facilitation.

When the mother is unable to retrieve her infant by picking it up herself, she resorts to what Harlow calls the "silly grin response" and the "affectional present" to lure the infant back within reaching range. "In 12 out of 22 observed occurrences, the silly grin was successful in bringing about an *immediate* return of the infant to its mother regardless of the infant's position or orientation." This interesting pattern is not yet reported from the wild since situations in which the mother cannot physically retrieve her infant are rare.

In rhesus macaques, the stage of maternal attachment and protection, which lasts about two months, closely correlates with the period of contrasting color and lack of facial communication. The lack itself, according to Rowell (1963), is a means of communication, which, like the difference in coat color, indicates a special status in the society: immunity from social controls that regulate all other social strata and the right to special protection. Special care, tolerance, and indulgence are integral features of that period of the infant's life. In fact, in at least one species, big males actually avoid dark infants and thus remove the most aggressive animals in the society from close association with the newborn. Male bonnet macaques appear to fear dark infants and quickly move away when an infant on wobbly legs approaches them. If it succeeds in reaching them, the infant climbs over, crawls under, and hops around them. The usual male reaction is to cringe and move away from the infant. Female bonnet macaques react favorably to dark infants, showing great interest in them and attempting to groom them.

The color differentiation of newborn infants tends to be more marked in species whose social relationships are characterized by the threat of aggression. Macaques and baboons require a clear signal to divert aggression from the newborn. Among chimpanzees, on the other hand, in whom the color differences are less marked, the infant gradually moves into an increasingly adult relationship with other members of its society with no more marked change than the slow and gradual darkening of skin color. There is ample evidence that this change (at two months for macaques and four to six months for baboons) thrusts the young infant into a new set of social relationships that require new responses from it. The infant begins by using facial communication more and more obviously and in turn, is subjected to disciplinary threats and attacks by the high-ranking adults.

That the color change and status change go hand in hand is confirmed by the difference between infant spider and howler monkeys in the length of their dependence upon the mother. The spider mother shows strong signs of maternal attachment and protection for several months longer than the howler mother and is still careful of her infant by the time the howler mother has become ambivalent. What is relevant here is that the infant spider monkey remains black for almost ten months before changing to the reddish adult color, whereas the infant howler changes from grey to adult black by its fifth or sixth month.

Another indication of the first stage of infancy, but of secondary importance, is the mother's reaction to the infant's movements. Although

PLATE 11   Japanese macaque (*Macaca fuscata*) female and her young black infant.

a mother will continue for a while to care for a dead infant, apparently movement is necessary to reinforce the bond between the two and to allow the relationship to develop. Immediately after the infant is born, the mother begins to lick off the afterbirth. When the licking gets too rough or when the afterbirth has been licked off its face, the infant turns its head away. Perhaps because of the lack of movement, caged mothers often continue to lick the faces of their stillborn infants until the eyes and parts of the face are gone. A small infant, wobbling unsteadily near its mother, may reach out and touch her, whereupon she picks it up and places it in her lap or under her belly if she is standing up. If the infant chirps and twitches excitably while walking, the mother immediately grabs it. A mother, feeling her infant slipping as it clings to her belly, will hitch it up and press it closely to her to give it a firmer grip.

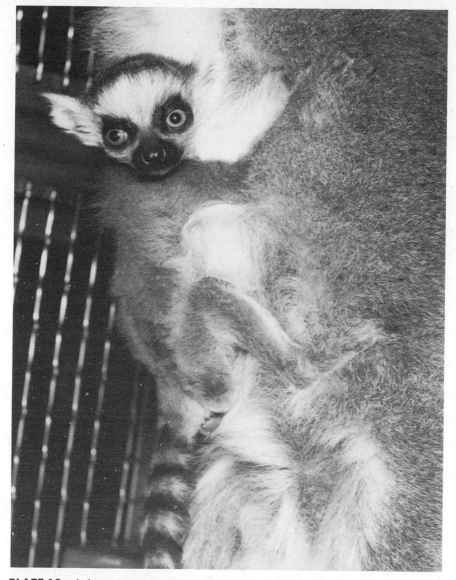

PLATE 12   Infant ring-tailed lemur (*Lemur catta*) clinging to its mother's fur. (*Courtesy of the Oregon Regional Primate Research Center; taken by Harry Wohlsein.*)

Free-ranging primates   With rare exceptions, the wild primate mother carries her young infant with her wherever she and the social group go. During the first stage of infancy, the infant is generally too weak and uncoordinated to walk more than a few shaky steps and is quite incapable of keeping up with the group when it moves. Therefore, the mother either holds it in her lap while sitting or carries it, usually under her belly, while moving. Despite her efforts to scoop her infant into a riding position when she begins to move and to assist it into a better riding position, often while running and leaping through the foliage she cannot give it much assistance.

The carrying of dead infants reported for several species in the wild indicates how deeply ingrained is the need to care for the infant. Van Lawick-Goodall (1967) reports that Mandy, a female chimpanzee, carried her dead infant for as long as three days, and Schaller (1963) saw a gorilla mother carry her dead infant four days before leaving it behind. Grey langurs sometimes carry their dead infants until they are decomposed (Jay, 1963), and savannah and hamadryas baboons have been seen with dead infants (DeVore, 1963; Kummer, 1968a). The extremely difficult feat of carrying a dead infant that cannot assist itself by clinging suggests that the mother's role here is not one of passive permissiveness.

Since neonate chimpanzees and gorillas are not so far advanced in their motor coordination as most monkeys, they require more support in the early weeks of their lives. As much as three months are needed for the infant chimpanzee to develop sufficient motor control for efficient and successful clinging. Thus, because of the infant's inability to support itself on her body as she moves, the chimpanzee mother must make a number of adjustments in her locomotor patterns and activities. "Most mothers, whilst moving about in the trees, carefully support their small infants either with one hand or, more usually, by exaggerated flexion of one thigh. During brachiation both thighs may be drawn up towards the body" (Van Lawick-Goodall, 1967). Some mothers walk in a semi-crouched position, which gives the infant support on the mother's thighs as she moves.

J. J. Petter (1962) reports that the Cheirogaleinae *Varecia* and *Hapalemur*, whose infants are relatively premature, deposit their newborn in a nest and carry them in their mouths when they wish to move them. Among nesting primates (excluding the great apes), the mother often carries the infant in the middorsal or midventral region or by the loose skin of the neck (Sauer, 1967). In *Galago senegalensis bradfieldi* this latter method was used less as the infant grew. Among marmosets, which probably use nests in the wild, the males carry the infants much of the time, transferring the infant to the mother only when it wishes to nurse. Through such care, these nesting forms also develop strong social bonds, the male playing a strong role in the social system.

Another major activity of the wild mother is to protect her infant from the unwanted attentions of other members of the group. These vary from species to species. Japanese and rhesus macaques allow the newborn infant only to be groomed and examined by other members of the troop, whereas langurs, both Hanuman and Nilagiri (Jay, 1963; Poirier, 1966), pass the neonate around among other members. Patas females are even more restrictive than the macaques and will not permit other patas to handle their infants (Hall, 1965, 1968b). Baboon mothers, however, permit other adult females to pick up and fondle their infants once they have displayed "greeting" behavior toward the mother (Hall, 1962b). Langurs pass their infants primarily to other adult females, whereas vervets (Struhsacker, 1967a) to a great extent permit juveniles and subadults to carry their infants. Occasionally, male savannah baboons and male Nilagiri langurs carry an infant (DeVore, 1963; Poirier, 1966), but hamadryas males are quite active in caring for infants in this first stage (Kummer, 1968). More than any other species, male mangabeys (Chalmers, 1968b) and marmosets (Hampton, Hampton, and Landwehr, 1966) are directly involved in caring for the newborn infants. Comparative studies of these different species show that the variations in male involvement with neonates have a long-range social significance. The adult's participation in the social milieu is directly related to its experiences as an infant.

Most of the mother's time in the initial weeks and months is spent in licking, grooming, examining, and watching her infant. The first few days the mother acts as though she has a new toy that must be examined carefully. Time and again, she will pick up the infant, turn it over, stretch out its limbs, and look it over carefully; grooming often accompanies this manipulation. Bonnet macaques treat female and male infants differently. If the baby is a female, the mother examines its genitalia a few times; but if the infant is a male, both the mother and other females examine its genitalia minutely and repeatedly.

Once the novelty begins to wear off, the mother's interest wanes somewhat, but it remains fairly strong until the color change. She continues her constant watch over the infant and retrieves it whenever it moves more than a few feet from her. The howler mother raises a loud hue and cry when her infant falls to the ground. Only the Nilagiri langur mother seems indifferent to her infant's mishaps in the trees; even if it cries out in a precarious situation, she comes to its aid only after it has fallen.

Stage I of infancy lasts from two to ten months among both Old and New World monkeys. Macaques change color at one to two months, baboons and langurs at about four to six months, and spider monkeys not until about ten months. Among gorillas and chimpanzees, the changes are so gradual that it is difficult to say when Stage I ends and Stage II begins.

Stage II—Maternal Ambivalence

According to Harlow *et al.*, the second stage is marked by maternal ambivalence. During this period, the mother still offers the infant protection, holds it, and is obviously attached to it. But more and more she is inclined to punish the infant and at times to reject it. Among laboratory rhesus, the punishment increases until the fifth month, when the increasing social distance between the mother and the infant removes it from her vicinity more and more frequently. Hence the incidence of punishment decreases, not because she is less willing to inflict it, but because the infant is well out of her reach!

Often during this stage there are temporary periods of rejection, but the mother usually returns to her infant between rejections and maintains fairly close maternal social relationships with it. A major disruption and the most usual one in many species occurs when the mother again enters into sexual activity. While she is in estrus and mating with males, she is likely to abandon her infant, which probably no longer nurses for sustenance and can therefore make its way quite satisfactorily within the group structure. After estrus, she again takes her infant to her breast, allows it to hold her nipple as a pacifier when it is frightened, and supports it when it is threatened.

In the laboratory studies (Harlow, Harlow, and Hansen, 1963), the stage of maternal ambivalence did not terminate because the normal social relationships with other monkeys were not available. In a typical social group consisting of other females, males, and young, the birth of a new infant would have terminated this stage and initiated the third stage. However, even in the wild some females do not give birth to a new infant each year. In this case, last year's offspring continues in the stage of maternal ambivalence, which then terminates slowly because of the infant's tendency to move more and more with play groups and because of the mother's slowly waning interest. In such circumstances, there is no sudden, traumatic displacement for the infant.

Free-ranging Primates Among free-ranging groups of primates the mother's ambivalence is probably a result of her partial return to more normal social relationships, the gradually increasing amount of time spent by the infant playing with its peers and juveniles, and the greater economic independence of the infant. The color change and rapid acquisition of a full repertoire of communication signals simply serve as obvious indications of the change.

The mother no longer carries the infant constantly during troop movement but allows it to move farther and farther away before retrieving it. She becomes increasingly concerned with other adults in the group, generally females.

Nursing is no longer the major source of nourishment for the infant, which by now has learned the major food sources of the group and can exploit them without the help of any other member of the society. Here it should be noted that in addition to obtaining nourishment, the infant primate also sucks its mother's nipple to experience comfort and security when it has been injured or frightened.

Nursing for sustenance slackens rapidly after the beginning of Stage II. The problem of observing its termination in the wild is that taking the nipple for comfort continues to be a major activity until the infant is forcefully weaned at the end of this stage. After the first few months, sucking movements of the mouth are rarely observed when the infant takes its mother's nipple; hence it is reasonable, I think, to assume that only the need for comfort is being met.

The forcible weaning of the infant usually takes place shortly before the birth of a new infant. Among many Old and New World species, the mother weans her infant by actively threatening and occasionally striking it. The process, which may last for several weeks, alternating between acceptance and rejection by the mother, sometimes is not finally terminated until the actual birth of the next infant. In fact, for most species it probably begins with the color change. In Stage I, the mother generally permits her infant to suckle whenever it desires. In Stage II, the deciding factor is the mother's willingness, and she becomes more and more impatient with the infant's attempts to nurse, finally resorting to the active slapping and threatening sessions just before full rejection.

Late in this stage, patas monkeys nip and slap their infants hard enough to cause them to squeal if they attempt to nurse (Hall, Boelkins, and Goswell, 1965). Vervets do much the same thing, but the vervet mother appears to be less successful than the bonnet macaques in keeping the infant away from her nipples. The bonnet macaque mother has been known to cause her infant to drop to a lower branch, screeching, by giving a mild stare threat. The vervet mother less forcefully removes her nipples from the infant by straightening her spine and pushing its head away and by occasionally slapping it. The infant vervet responds to the weaning by ceasing to suckle, but it may or may not leave its mother (Struhsacker, 1967b).

According to the field reports, the onset of weaning varies considerably. Struhsacker (1967b) states that weaning can begin as early as twenty days among vervets. Bonnet macaque mothers begin in earnest around nine months of age, but savannah baboons and langurs (DeVore, 1963; Jay, 1963) wean their infants between the eleventh and fifteenth month. Unlike most primates, chimpanzee mothers apparently do not wean their infants but permit them to nurse as long as they show any interest (Van Lawick-Goodall, 1967). Not so tolerant, gorillas deliver an occasional mild rebuff (Schaller, 1963).

During the stage of maternal ambivalence, the mother primate helps her infant in other ways. Since the infant no longer rides its mother regularly but is still not very successful in locomotion, the mother helps it to move around in the trees. Among howler monkeys, chimpanzees, and gibbons, for example, when there is a space too wide for the infant to cross, the mother pulls the branches together to form a "bridge."

Among the many primate species that give birth every year, the second six months of the infant's life is also the mating season for the mother. When the infant bonnet macaque is six months old, its mother reenters the sexual cycle by coming into estrus. At the first, and sometimes the second, estrous period, the bonnet mother deserts her infant to consort with the males in the troop. But once the peak of estrus is over, she returns to her infant and resumes the relationship. In fact, even during estrus, she probably does not prevent it from sleeping with her in the large trees during the night and only rejects it during the daytime periods of sexual activity which, at the peak of estrus, may take up most of the day. In later estrous periods, when the infant is approaching nine months to one year of age, copulations occur from time to time but do not seem to have the same disruptive effect upon the mother-infant relationship as the earlier ones. In fact, the mother may be pregnant during later copulations. In such cases the infant is not rejected or abandoned, but since it is older and spends most of its waking hours with its peers, it spends less time with its mother.

In this second stage, the hamadryas baboon mother is less concerned for her infant than other higher primates. She no longer retrieves her infant as she did when it was black, and the infant is allowed to move off to other harem units if it so desires. As we will discuss in the chapter on social organization, the hamadryas infant has other opportunities for close social ties that are relatively weak for most primate species. I suggest that these differences in social opportunities are directly related to the social adaptation of the hamadryas baboons that use the male harem groups in dry country.

For macaques and howler monkeys, Stage II ends with the birth of the new infant. But for langurs and baboons, the close bond is terminated by the beginning of the mother's estrous cycling. Chimpanzees, on the other hand, seem to continue the close relationship with slight diminution even after the birth of a new infant. However, for all species there is evidence that the tie between mother and child is never completely severed while both live.

Stage III—Maternal Separation or Rejection

Harlow's final stage is that of maternal separation or rejection. With the birth of a new infant, the mother begins the cycle all over again, concentrating her whole attention now on it. The previous year's infant is no longer permitted to sleep in the close ventral position with its mother. It is groomed considerably less than before by its mother, which no longer gives the new juvenile regular support in its social encounters although, depending on her rank and proximity to the encounter, she usually offers occasional support.

This rejection of the infant by its mother is probably more traumatic an experience than the process of weaning. Here the infant ceases nursing for sustenance before the mother is pregnant again and, in many species, the amount of time the infant may take the nipple for comfort is only gradually curtailed. But in rejection last year's infant is suddenly and markedly bereft since all of the mother's care and attention now concentrate on the new infant in a renewal of maternal attachment and protection. Last year's offspring must fend for itself, within the society, literally overnight. Now it is deprived of the comfort and warmth of the mother's body, and it can no longer sleep in her lap, huddled against her chest.

However, as mentioned earlier, rejection does not come at the same time for all primates. In baboon and Hanuman langur society (DeVore, 1963; Jay, 1963), rejection occurs when the mother reenters sexual activity and becomes pregnant. However, baboon mothers space their infants more widely than macaques, and whereas the rejection of the infant rhesus at the onset of the first postpartum estrus is temporary, that of the infant baboon is more permanent. It should be recalled that the age of the rejected offspring, both developmentally and actually, is about the same for baboons, Hanuman langurs, and macaques. When the macaque mother begins her postpartum estrous cycling, the young macaque is still too immature to be deprived of her ministrations, but the greater spacing of Hanuman langur and baboon infants makes it possible for the rejected infants to fend for themselves.

The chimpanzee mother, on the other hand, never reaches a point in her relationship with her offspring that could be called rejection. According to Van Lawick-Goodall (1967), the mother does not play an aggressive role in weaning her young. The drying up of the mother's milk appears to be the factor that normally terminates suckling. Although the lack of milk is upsetting to the infants themselves, the mother neither actively nor passively prevents her infant from making contact with her nipples. Thus, despite the cessation of milk flow and suckling, the mother–infant relationship changes but little. The relaxed continuation of this relationship is also indicated by the ability of the infant to take food from its mother; this is also reported by Schaller (1963) for gorillas. Since neither of these species is markedly aggressive in its social relationships, the remarkably abrupt character of maternal rejection among rhesus and a number of other species may be a factor in producing the aggressive adult. Or, perhaps, the other way around: maternal rejection is the result of an aggressive society and hence does not occur in primate societies where aggression does not exist or exists only slightly.

Among hamadryas baboons, the rejection is probably not so traumatic for the offspring, which in most cases is taken into the care of a young male. Kummer (1968) believes that the male drive to care for young females is at least in part related to the maternal drive. The care given by the male is quite similar to that of the mother. In the later stages, the mother hamadryas shows considerably less interest in her infant than other primate mothers.

THE INFANT'S DEVELOPING RELATIONSHIP WITH ITS MOTHER

Both field and laboratory studies indicate that in the mother–infant dyad the infant primate is not merely the passive subject of the mother's attention. From birth onward, the infant primate increasingly affects the development of its relationship with its mother. Because it is more mature at birth than the human infant, its motor coordination and reflexes are better developed. In many species, the infant can cling to the mother's abdominal fur almost immediately after birth and root for the nipple in order to suckle. This ability to cling and even to move about somewhat on the mother's body enables it to determine the closeness of their relationship to a degree impossible among humans. Some primate mothers that do not wish to have anything to do with their infants manage to completely avoid the relationship; but if the mother is at all permissive, the infant has the necessary reflexes and motor abilities to take an active part in their relationship.

The development of the infant's affectional system for its mother has been nicely elucidated by the studies of Harlow, Zimmerman, and their associates. In the first of a series of experiments to isolate the variables that determine the development of that system, they used eight infant rhesus monkeys that had been separated from their mothers within six to twelve hours after birth. All of these infants were raised individually with two surrogate mothers, one a cloth-covered mother, the other a bare wire form. As a test of "the relative importance of the variables of contact comfort and nursing comfort," four of the infants nursed on their wire mother and four on their cloth mother (Harlow and Zimmerman, 1958).

A timing mechanism to record the amount of time spent on each surrogate mother indicated that whether the infants were nursing on the cloth or the wire mothers they spent up to seventeen hours a day on the cloth mothers and only one to two hours a day on the wire mothers. Thus, the primary variable that affects the infant–mother relationship is the infant's ability to cling to her furry body rather than the mother's role as a source of nourishment.

Further experiments, designed to show the relative importance of clinging and nursing, used a pair of cloth surrogate mothers of different colors, one of which was a lactating mother and the other not. With the urge to cling satisfied by both mothers, the infant chose to spend more time on the nursing mother than on the nonnursing mother.

With these and other experiments as a basis, Harlow (1960b, 1965) describes four stages in the development of the infant's affection for its mother: reflex, attachment, security, and separation.

The Reflex Stage

The reflex stage lasts only a matter of days (between ten and twenty for rhesus infants) and has been most clearly observed in the laboratory setting, where the infants can be purposefully manipulated. It is characterized chiefly by two groups of reflex reactions to stimuli: one is associated with nursing, the other with close physical contact (Harlow and Harlow, 1965). A third reflex, the righting reflex, is subordinate to the others. Harlow and Harlow demonstrated this subordination to clinging by placing an infant rhesus monkey on its back on a flat, solid surface. The infant immediately rotated to a prone position. But when it was placed on its back while clinging to a cylinder, it continued to lie passively without attempting to right itself.

The nursing reflexes, which are present in all primates whenever laboratory studies or field observations have been adequate, need not concern us greatly at this point. The rooting reflex (the infant's search for the nipple) is basic to nursing and to the infant's survival. The involuntary reflex seen in infant monkeys and apes causes them to climb upward and is probably related to the nursing complex. Since the nipples are pectorally located in most primates, the climbing reflex brings the infant into a position where the rooting reflex will be effective in locating the nipples. Prechtl (1958) and Hooker (1952, 1954) discuss the rooting reflex in man. Hooker found response to tactile sensation around the mouth of a human fetus at seven and a half to nine and a half weeks of menstrual age. The response increases in intensity with age and becomes localized rather than generalized after the thirteenth week. Usually in man the response is dormant until a few hours after birth, but in other species more advanced in their motor development it is functional immediately after birth.

The clinging and grasping reflexes are associated with close physical contact. The former enables the newborn monkey to maintain close contact with the mother's abdomen; the latter, which involves the inversion of the foot, gives the firm and secure grip essential to maintain that contact. Between the tenth and twentieth day in newborn rhesus monkeys, the grasping reflex is suppressed cephalocaudally (from head to tail) and is replaced by coordinated motor control of the limbs whereby the infant can break and regain contact with its mother. Thus for the first time the element of choice becomes a feature of the infant—mother relationship. Thereafter learned behavior takes precedence over reflex behavior, and the infant soon enriches its behavioral capacities through learning.

Bolwig's observations on a growing patas monkey support those of Harlow *et al.* The infant gripped any object that was placed in his hand and screamed loudly if it were removed, "reacting more strongly if an object was removed from his hand than from his foot" (Bolwig, 1963). This latter observation suggests a grasping reflex similar to that of rhesus monkeys and the cephalocaudal suppression of reflex. Bolwig contends that the reflex, clinging, and exploratory behavior cannot be isolated into separate, progressively higher stages of the infant's life but rather occur together with one or another phase predominating at any given time. Thus the stages should probably be considered as aspects of the developing infant's behavior rather than discrete segments of a sequence in time.

Free-ranging Primates Reflex behavior in free-ranging primates is not easy to demonstrate since the animals are usually not available for manipulation at close quarters. However, there are indications that the clinging and grasping reflexes have evolved in response to at least one strong selective factor. In all but a few species, the newborn monkey or ape comes into a society that does not maintain a home base where sick and helpless individuals can be left while the rest of the troop forages for food. The group moves daily throughout large parts of its home range, and from the day of its birth the infant must travel with its mother in the troop. To free her hands to climb in the trees and jump from branch to branch, the infant must be able to cling tightly to her. While thus engaged in arboreal acrobatics, the mother has little opportunity to aid the infant, which without a strong grasping reflex would probably be dislodged and dropped to the ground many meters below. Although the mother can pause from time to time to press her infant more tightly against her belly, most of the support must come from the infant alone.

There are some exceptions to this general pattern. Several genera of prosimians nest in hollow trees or leave their infants in the fork of a tree or plant while they forage for food. According to Petter (1965), *Hapalemur griseus, Lemur variegatus,* and the Cheirogaleinae have multiple births and leave the young in a nest. To carry her young, the mother takes them into her mouth rather than allowing them to cling to her belly. Thus, if tested, these forms would probably not show strong grasping and clinging reflexes. In fact, in these and other such prosimians, a "relaxing reflex" can be shown to overshadow grasping and clinging reflexes. According to Sauer (1967), when a mother *Galago senegalensis bradfieldi* grasps her infant by a fold of neck skin or by the skin of the back or side during the first two weeks, the infant immediately relaxes and hangs tensionless. But as the infant grows older, it struggles against the carrying posture, and the mother must induce it to accept it. Apparently, an initial reflex soon disappears and must be replaced by learned behavior, just as the clinging and grasping reflexes of rhesus infants are replaced in the second and third weeks of life.

Some forms of prosimians, for example, caged *Galago c. crassicaudatus,* seem to adopt a "compromise" pattern, the infant clinging to the mother's belly in relaxed settings but being carried in her mouth when danger threatens (Buettner-Janusch, 1964). This seems to support my contention that in these forms a relaxing reflex supersedes clinging and grasping reflexes. When the mother is in danger, the safest carrying technique is by mouth. On the other hand, when the *Lepilemur mustelinus* mother must carry her newborn infant, she does so in her mouth even though it can progress unassisted on the branches (Petter, 1965). In the newborn *L. mustelinus,* the righting reflex probably takes precedence over the clinging and grasping reflexes.

Among many of the other Madagascar prosimians, including several species of *Lemur,* the Indridae, and *Daubentonia,* the newborn clings to the mother's abdominal hair in the first hours and days of life. It may shift to her back later on but only after the reflex phase has passed. The infant is occasionally given a helping hand and pressed closer to the mother's abdomen to obtain a better grip, but the mother and infant move with the group, the mother making the same leaps through the trees as the other adults, seemingly unhindered and undaunted.

All the evidence for Old and New World monkeys indicates that the clinging and grasping reflexes of the newborn infant must be strongly developed for it to survive. Many species in Africa, Asia, and South

PLATE 13 The thick-tailed galago (*Galago crassicaudatus*). (*Courtesy of the Oregon Regional Primate Research Center; taken by Harry Wohlsein.*)

America spend all or most of their lives high in the trees, moving from tree to tree by climbing across slender branches stretching between the trees or by a series of leaps across gaps, a few to many meters wide. When predators or suspected danger approaches a group of forest-dwelling Hanuman langurs, the troop makes a series of leaps from tree to tree to outdistance the enemy. During these escape maneuvers, all four limbs of the mother are needed for locomotion. Hence it is precisely then, when the going is roughest and the mother is least able to assist, that the infant must cling on, literally for its life.

Some species, like most baboons and many macaques, spend much time on the ground. These terrestrial species have less need, perhaps, of the strong clinging reflexes, but even they move into the trees for at least occasional feeding and for sleeping or protection from predators. Thus, although a fall under these circumstances would not be fatal, an infant that could not cling securely would hinder its mother's movement and be in greater danger precisely when the mother, running or climbing out of reach of a predator, is least able to give assistance.

Among the higher primates, only the chimpanzee, gorilla, and man seem to lack strong clinging reflexes at birth. The chimpanzee mother must give major assistance to her infant during the first month and the gorilla infants' grasping reflex is "so weak that they are unable to hang onto their mothers unsupported for more than a few seconds" (Schaller, 1963). Schaller (1965) described an infant gorilla that lacked the strength to grasp its mother's hair securely and had to be supported by one or both of the mother's arms until it was nearly three months old. These apes, mainly terrestrial (especially the gorilla), are large enough to handle almost all predators and thus rarely need to run in precipitate flight. Gibbons, a truly arboreal form, seem at birth to need little assistance. The highly arboreal orangutan also has a respectably developed grasping reflex at birth.

The Attachment Stage

As the reflexes are replaced by cerebral control of grasping, clinging, and feeding, the infant moves into the second, or attachment, stage of its relationship with its mother. Both the reflex and attachment stages occur during the maternal stage of attachment and protection. Among the infant rhesus monkeys used by Harlow *et al.* (1965), this stage lasted from sixty to eighty days after birth. During this time, the infant appeared to derive only a moderate sense of security from the mother's protection and seemed only transiently insecure during her absence. Their studies with surrogate mothers indicate strongly that at this time the infant's association with its mother is crucial for the development of normal social relations later in life.

During this period, the infant not only forms a strong attachment to its mother but is also able to draw strength from her presence in new and frightening circumstances. According to Mason (1965a), the presence of the mother serves to prevent the development of stereotyped behavior such as self-clasping, digit-sucking, and rocking. Once firmly established, this behavior later prevents the animal from forming normal social bonds. In its stead, the protected infant clinging to its mother has her fur to grasp, her nipple to suck, and the rocking motion of her movements. Such an infant has little need to seek security. In fact, the mother is so concerned with its security that it has little chance to escape her attentions! As a result, much of an infant's time not spent in nursing, sleeping, or being carried is spent in attempting to observe and manipulate its environment without the mother's interference. However, so successful is the mother's vigilance that the infant's brief escapes usually end ignominiously in its being held by the tail, pulled back to the mother, and clasped to her chest!

The infant's chief activities during this stage are riding or sitting on the mother's ventral flexure, nursing, and sleeping. But increasingly as the infant grows older, it spends its time in wobbly attempts to move about in the trees or on the ground, while the mother rests nearby. Generally the infant plays by itself. Even two infants playing close to each other do not seem to coordinate their efforts, probably because it is basically a period for testing and improving motor coordination.

The Security Stage

The security stage for the infant corresponds to the ambivalent stage for the mother. Now the roles are reversed. No longer does the mother continuously proffer the security of her single-minded attention. She has begun to revert to the wider relationships with other members of the society. Now it is the infant that must seek out the security of her body and assistance when it needs it rather than take the security for granted.

Partly because the mother is more relaxed and partly because the infant is more venturesome, the latter encounters more situations that cause it alarm. When it returns to its mother to huddle in her fur, take her nipple into its mouth, and look back at the terrifying object, the infant has the enviable opportunity to examine the object of its fear from a safe vantage point. Having done so, almost leisurely, it later returns to examine and study the object with new courage and with the knowledge that a safe haven is nearby.

According to Harlow and Harlow (1965), the security stage functions only if the infant has experienced a previous attachment to a furry mother. Their experiments involve infants that had been raised on either wire surrogate mothers or cloth surrogate mothers but not both. The latter group were placed with their cloth mothers in novel situations with new objects. Immediately the infants rushed and clung to their cloth mothers, turning after a few moments to examine their surroundings. Even after they had left the safe refuge, they returned from time to time as if to renew their courage.

On the other hand, when infants raised on wire mothers were placed in new situations with novel stimuli, they crouched in terror and made no attempt to examine the new objects and surroundings. Nor did they make any effort to contact their wire mothers and thus gained no emotional support from them. Infants raised on cloth mothers reacted with similar fear if the mother were not placed in the test situation with them.

Field studies likewise demonstrate that all infant monkeys and apes known at present return often and regularly to their mothers. The heavy reliance of primates on learned behavior requires that infants possess a mechanism for reducing fear to a level that will allow them to explore and examine their surroundings. Fear itself is a stimulant to exploration (Halliday, 1966) but the extreme fear of a motherless rhesus infant immobilizes it and deters exploration. On the other hand, in a normal social situation, fear arouses the infant primate to activity and prompts it to seek shelter with its mother. The comfort and safety of the mother buffers the fear, lowering it to a level that encourages a healthy examination and exploration while it maintains the necessary precautions.

The environment of free-ranging primates is so exceedingly rich that by comparison "rich" laboratory environments are impoverished. During the attachment stage of its development, the infant has enough freedom, through its mother's ambivalence, to move about in this habitat, exploring

new objects and making new social contacts. Moreover, the mother acts
as a base of operations for excursions into the environment to investigate,
to play, and to feed. The infant always maintains its contact with its base,
and any sign of danger (real or imagined) brings it quickly back. Only
as it grows older and gains the confidence that comes from knowledge
of the surroundings is the importance of this base of operations significantly reduced. Still, the group as a whole or subdivisions within the
group continue to play an important role in the security of the growing
and adult primate.

The Separation Stage

This stage begins with the mother's rejection of the infant, either at the
onset of postpartum estrous cycling, as with baboons and Hanuman langurs, or at the birth of a new infant, as with macaques, howlers, vervets,
and many others. By this time, the wild infant has established a series of
relationships outside the mother–infant dyad that are strong enough to
sustain it and bind it to the society as a whole. The primate mother's
rejection is rarely complete although among gibbons the mother eventually participates in forcing the offspring out of the family group (Ellefson,
1968).

Rejection–separation simply means that the infant turns more strongly
to other relationships outside the dyad but maintains close ties with its
mother. Because the mother-lineage is the basic unit of chimpanzee society, the infant chimpanzee shows the least rejection and separation from
its mother; it still spends many years in close contact with her. The hamadryas baboon infant, on the other hand, is in the process of establishing
new relationships as a prelude to forming a harem association, particularly the female infant, which is soon taken into a harem unit and replaces
the mother–infant bond with a male-female bond that seems to be
strongly similar to the former. But for the majority of primates living
in a troop organization of some sort, the mother–infant bond is replaced
or balanced to a greater or lesser extent by other social ties within the
group. This new equilibrium between social bonds determines the operation of the society into which the infant grows.

PATERNAL CARE

The very nature of sexual relationships in primate society ensures the
difficulty if not the impossibility of determining physiological paternity.
Of the higher primates, only among gibbons and *Callicebus* can one be
reasonably sure that the male of the group is the father of the offspring.
Paternal care, in the strict sense, is a misnomer. But if the males of the
group are seen to fulfill a generalized paternal role, then the relationship
an adult male has with an infant can be called paternal care. Here the

distinction is between physiological paternity and social paternity made in many human societies.

There are no laboratory studies on the stages of development in male–infant dyads comparable to those on females, but numerous good field reports show that paternal care is an important aspect in the early life of many primate infants. For a number of species this importance extends far beyond the general role of protector and procreator, and among hamadryas baboons it could well be the primary bond for the formation of new subunits.

As far as timing is concerned, paternal care among primates shows no cross-specific pattern. Itani (1959) reports that among Japanese macaques most of the paternal care is demonstrated during the delivery season, when last year's infants are being rejected by mothers about to give birth. I have seen male bonnet macaques hold infants for a short time around the beginning of the mating season when sexual behavior begins to disrupt the mother–infant dyad temporarily. Barbary apes (*Macaca sylvana*), East African mangabeys, and South American marmosets also show marked paternal care as the birth season approaches.

Generally, except for nursing, paternal care has the same characteristics as maternal care. The male carries the infant in the same position as the female, retrieves it when it tries to move too far away, grooms it, and protects it from the excessive attentions of others.

Chalmers (1968b) reports that in a troop of seventeen mangabeys two of the four adult males associated noticeably with the newborn infants. This association began during the tenth week of the infants' lives, the period of intensive maternal care, and reached a peak during the fourth and fifth months, during which the infants spent much more time with the males than with their mothers. At the peak of this male–infant association, the third of the four adult males was also seen carrying an infant once. Except for nursing, paternal care among the mangabeys is equivalent to maternal care. The infant responds to the male as it would to its own mother and therefore has double protection and attention. However, it should be noted that Chalmers has reported another mangabey group that apparently lacked paternal care.

Lahiri and Southwick (1966) observed that among barbary apes a strong positive social bond exists from the beginning between the dominant male and the infants. The male barbary ape uses three methods to acquire the infant from the mother. He may simply walk up and take the infant from the unprotesting mother or give a submissive or soliciting gesture such as lip-smack and teeth-chatter before taking the infant. In this second method, the mother sometimes holds onto the infant but more often lets it go; or the infant may go to the male of its own accord. Finally, the male lies prone in front of the mother or infant and resorts to the same gestures; here the infant invariably leaves the mother for the male's back.

The male marmoset (*Oedipomidas oedipus*) often does most of the

TABLE 14.   Parental Care in *Macaca sylvana*

Infant's Age	Time Spent With Mother	Time Spent With Dominant Male	Time Spent With Other Adults and Juveniles	Time Spent in Independent Play
0–4 weeks	82%	7.5%	5.7%	5.6%
4–8 weeks	72.9%	7.6%	1.9%	17.6%
8–12 weeks	51.5%	8.1%	0.9%	39.6%

Lahiri and Southwick, 1966.

carrying and caring for the infant, turning it over to the mother only to nurse. Such care, however, has been observed only in laboratory situations.

Bonnet macaques do not lavish this kind of paternal care on their newborn infants, but just before the mating season, some adult males have been observed to carry and watch over four- to six-month-old infants. Only Kaufman and Rosenblum (reported in Harlow and Harlow, 1965) have cited the example of a male bonnet macaque adopting and caring for an infant.

Itani's (1959) summary of paternal care among Japanese macaques is the most extensive that has been reported. During the delivery season a number of year-old infants are rejected. However, typically the second year of a macaque's life is the most dangerous period. No longer the focus of its mother's attention, the infant is not large enough in some respects to care for itself effectively. Before rejection, it probably slept every night warm and safe against its mother's chest. Now it huddles haphazardly against another monkey's back for warmth at night. The dominant Japanese macaque male tends to fill the void left by the mother's rejection. Sometimes the male is as solicitous as the mother even in the first months of the infant's life. He carries the infant and rescues it from difficult situations, thus tiding it over the difficult period of adjustment to its mother's rejection. Later, as she becomes more ambivalent toward her new infant, she renews her relationship with her older children but never as closely as before.

Sometimes this paternal care is tendered not by the high-ranking males but by lower-ranking males, which seem to use their "altruistic" behavior as a bid for closer relations with the higher-ranking animals. While caring for an infant, they are accorded higher status and are permitted to move closer to the center of the troop than otherwise.

Among hamadryas baboons, which favor the harem system, the situation is unique. Although the harem leader often carries dark-colored infants, he ceases to do so at the color change (four to six months) at the same time that the mother ceases to retrieve the infant. Thus the bond between the hamadryas infant and its parents weakens remarkably after the color change. This opens the way for "paternal behavior" of another sort, since the rejected infant is cared for by subadult males outside the parental harem.

As he achieves young adulthood, the male hamadryas tries to kidnap infants from their mother. Generally he is unsuccessful while the infant is young enough to depend upon its mother as a source of food, and the mother with the help of the harem leader is able to keep her infant. After the infant's color change, the chances increase that the outside male will succeed in keeping the infant. Most often it is the subadult males that attempt to kidnap black infants, whereas young adult males tend to adopt older infants and juveniles. In fact, Kummer (1968a) noted that when a hamadryas mother dies and leaves an infant, the latter is generally cared for not by the unit leader or another female, as so often happens in other monkey societies, but by one of the young males, who tends thereby to establish a prolonged relationship. Kummer feels that this adoption through kidnapping may ease the trauma of maternal rejection for the juvenile since the latter can turn to its adopter for the same kind of attention given it by its mother.

Only in the biological priority of birth and nursing does the hamadryas mother–offspring supersede the adult male–young female adoptive relationship. Very often the infant female is adopted by the male and ultimately becomes the basis of a new harem unit. During the early months of this relationship, which resembles the mother–infant relationship, the infant or juvenile female runs and clings to the male when there is danger or alarm. The male occasionally carries his young consort on his back much as her mother had done and also helps her climb over difficult terrain. The female even uses the same vocalization, the soft hum, when left behind momentarily, and the male responds as the mother would, returning to pick her up or retrieve her.

Among savannah baboons, paternal care is less institutionalized. Males, however, do carry infants and are very conscious of their presence (De-Vore, 1963). Males have been reported by more than one observer dropping back to accompany (and hurry?) a female with a young infant who has lagged behind the troop.

A number of other primates show no special paternal care beyond the protection afforded the mother and her offspring by big males which attack any other animal that threatens them. Bowden, Winter, and Ploog (1967) noted that adult male squirrel monkeys show little interest in the newborn infant. However, squirrel monkey males tend to be peripheral to the group's social activity. Southwick, Beg, and Sidiqqi (1965) reported that the adult male–infant contacts in the Aligarh temple groups of rhesus macaques were aggressive in nature, the males attacking the infants if they took notice of them. But such aggression is probably confined to the overcrowded temple situation since the adult male rhesus on

Cayo Santiago showed some positive, though rare, social bonds between themselves and the infants together with occasional aggressive attacks on the latter. Even on that island they are overcrowded.

The male patas monkey shows none of the protective behavior characteristic of baboons when the infant is threatened. In fact, the male patas is rarely in close proximity to the female and her infant, his role being that of a decoy rather than of an armed bodyguard (Hall, 1965, 1968b).

With such a range of variation in "paternal care," it is obviously difficult to generalize about this behavior among primates. The general male protection of females with young infants is occasionally elaborated for special purposes, for example, to assist at particularly vulnerable periods in the life of the infant or to strengthen the social bonds of the group.

WIDER RELATIONSHIPS

Even in the smallest of primate social groups, the mother–infant dyad is embedded in a complex matrix of social relationships. The gibbon family usually includes siblings as well as mother and father. In a bonnet macaque troop of fifty, the complexity is staggering; and in those species like some Japanese macaques and herds of hamadryas baboons that go into the hundreds, the complications become so difficult for the individuals to handle that major subdivisions become mandatory.

Primates have evolved the social group into a survival mechanism of such importance that the individual is particularly vulnerable without it. In fact, so great is the dependence on the social group that infants raised in isolation can hardly be considered primates in the social sense. Many laboratory reports show how unadaptable the isolated primate is.

Females raised in captivity and isolation tend to be inept mothers. Captive chimpanzee mothers are often afraid of their first-born, not only refusing to allow the infant to cling to them or touch them but also avoiding handling the infants themselves (Yerkes, 1943). Van Den Burghe (1959) has reported a captive gorilla mother who, upon pulling her newborn infant from her vagina, proceeded first to bite off a hand, then a foot and, finally, to puncture its skull with her teeth. Harlow's experience with mothers raised in isolation lends further support to this generalization. The isolated rhesus mother's rejection of her newborn infant, despite its repeated attempts to make contact with her, dramatizes such unadaptive behavior, which contrasts with the careful protectiveness of the wild mother.

Such unadaptability is not reported among wild groups under usual conditions. The mother has ample opportunity to see other mothers caring for their infants before she has her own. The first attempts at caring for her own infant are likely to be inefficient compared with those of a multiparous mother, but they are nonetheless usually successful in that the infant is given basic maternal care. The fact that some infants may die from disease, overexposure, or predation is perhaps partly due to some deficiency in parental care but not to a direct assault or rejection by the mothers.

In fact, a number of fairly obvious behavioral mechanisms operate to ensure that the maturing female knows what an infant is and how to handle it. The most dramatic is that of Hanuman and Nilagiri langur females who, within a few hours after the birth of their infants, hand them around· to other females in the troop, including subadults. The comfortable infant rests peacefully, in the new female's grasp, but the uncomfortable infant squirms and complains and is retrieved by its mother. This pattern is not confined to langurs. Among vervets, usually juveniles and subadults handle the infants and thus accustom themselves to them. *Cebus albifrons* females in the New World also carry infants other than their own, even in the nursing position (Bernstein, 1965). Among baboons, the infant of two months is often grasped by other females and clasped to their bodies; however, the infant quickly disengages itself and returns to its mother (DeVore, 1963).

Commonly, the newborn infant and its mother become the focus of attention for other troop members. Members of a *Propithecus verreauxi verreauxi* troop gather around and attempt to groom the newborn infant for a number of weeks. The infant *Lemur catta* is a focus of attention for the other females but not for the adult males; in fact, the mother permits only other mothers to groom her newborn (Jolly, 1966).

Most macaque females will not permit their infants to be passed around but do allow some examination and grooming as long as the infant remains in their grasp. Bonnet macaque juvenile and subadult females are often seen close to a mother–newborn infant dyad, either attempting to groom or simply looking on. Carpenter (1934, 1935) reported that howler mothers also are the center of such attention.

PEERS

In most social groups, the infant has an opportunity to associate with other young primates. In the larger troops, age sets often form play groups and thereby a secondary center of activity for the infant after its first few months. Since the major activity of these age sets and other peer groups is play, this aspect of social relations is discussed in Chapter 9 under that heading.

Chapter 6
Communication

Although all animals depend for survival on sensory information received from the environment, the dependence varies from species to species and from one sense to another. Some animals depend mainly on the chemical signals received by their olfactory and gustatory senses, some on the mechanical signals received by their tactile sense, and still others on the vibratory and optical signals received by their auditory and visual senses, respectively. In addition, some animals have electrical transmitters and receptors. In primates, the senses of touch and sight are the main channels of information about the environment.

Generally speaking, all stimuli communicate some information about the environment. But in this chapter, we are limiting communication to mean only those signals that influence the social behavior of primates. Recently Cullen (1972) described animal communication as signals that pass between social animals and that "help to mold each other's behavior toward some goal which is to their mutual advantage." From a slightly different point of view, Haldane (1955) had stated that "animal X communicates with animal Y if it produces a signal describable in the language of physics or chemistry which alters Y's behavior." Note that in restricting communication to signals that influence the behavior of other animals, these authors eliminated the broad range of signals received from the environment: vegetation, weather, sunlight, and other aspects of the inorganic environment. Although they did not eliminate signals received from other species of animals, they focused on intraspecific communication and thereby emphasized signals that influence behavior. Among feral animals, communication in this sense is limited almost entirely to members of a social group; only occasionally does it occur between members of neighboring conspecific groups and rarely between members of different species.

Such communicative signaling is not necessarily deliberate. For many species, a number of communicative acts are triggered by innate responses to the internal or external environment. It is unlikely, therefore, that these animals consciously emit signals to influence other animals around them. On the other hand, animals that have evolved a capacity for learned behavior—and primates have done this to a high degree—can consciously manipulate signals. Mammals generally can alter their behavior patterns through learning and display a flexibility of response impossible in animals whose behavior patterns are largely determined by inheritance.

Groups of social beings require communication signals to ease the flow of social interaction and to coordinate group activities. The initial movements of some actions—for example, reaching for other individuals or lunging at them—have been elaborated and ritualized into special signals that impart social information that goes beyond the mere fact that a group of conspecifics is present. Not only is communication necessary for social cooperation but Bastian (1968) has argued that social behavior and animal communication are one and the same. There can be no social behavior without communication, and communication between two individuals must necessarily be social.

PRIMATE MESSAGE MEDIA

When the ideas and premises of communication theory are applied to the study of subhuman primates, several different communication media emerge. Haldane (1955) listed four different signal media used by subhuman primates: chemical, optical, vibratory, and kinesthetic (tactile). All four share two functional attributes that make them suitable for communication: each contains information and each plays a role in helping the communicating primates to adapt.

Chemical Signals

Chemical signals, or *pheromones* as they have come to be called, involve the senses of smell and taste. An outstanding feature of the chemical channel is that it provides a long-lasting signal that can be left behind, like a chemical deposit on a tree branch, or can be emitted continuously while the animal is in a particular physiological state, like a female in estrus. The advantage of such pheromones is that they do not reveal the transmitting animal's precise location to a possible predator but convey information to a conspecific in the general area.

Several such messages are transmitted by various primates. One is territorial marking, that is, the deposition of scent on various branches in the area occupied by the group. A. Jolly (1966) noted that *Propithecus verreauxi* mark branches that are later sniffed and marked by members of other groups. A second message is the physiological change in an estrous female. In many species, females in estrus emit a distinctive odor that is recognized by males, which respond accordingly. Information about food is, perhaps unwillingly, transmitted by individuals that return from a food source with full mouths and find themselves being sniffed in the face by their fellows. But the olfactory sense conveys only a minimum of information in primate communication. The disadvantages of chemical signals are the difficulty of directing messages to a particular individual and of sending sequential signals when the earlier ones persist and interfere with the message of later ones.

Optical Signals

Optical signals play a major role in subhuman primate communication. The advantages of the optical channel lie in the large number of discrete messages that can be transmitted, the directionality of the messages, the partial broadcast reception, and the rapid fading of the signal.

The facial and bodily gestures used by primates to convey messages are many and varied. The facial gestures, at least, can be directed to a particular recipient but at the same time can be perceived by other members of the group that, once the message is understood, can make the proper social adjustments to the interaction. Since these optical signals fade rapidly, there is no possibility that successive signals can be confused.

The content of optical signals in subhuman primate communication most often includes information about the relative status of individuals in the society, mediation of social conflict through threat and subordination messages, the receptivity of estrous females and the coordination of mating behavior, the recognition of species, and much of the mother–infant interaction.

Vibratory Signals

Vibratory or vocal–auditory signals, the primary medium of human language, are secondary in subhuman primate communication. Only a few vibratory messages stand alone as complete units; for example, the warning call that precedes the approach of danger is not accompanied by any other reinforcing signal. But most calls among Old World monkeys and apes, such as the screeches or growls accompanying a threat, are used to emphasize the basic optical message or to signal the recipient and to ensure that he gets the message. New World monkeys have more varied vocalizations, and more of their vibratory signals may be meaningful utterances in their own right.

Vibratory signals are particularly advantageous when the animals are moving through thick foliage. The signal is broadcast in every direction and is heard by everyone within range. Once given, the warning call need not be passed from individual to individual; all individuals can react immediately and appropriately to it. Because of directional hearing, the sender of a vibratory signal can be located fairly easily and can draw the receiver's attention to signals in other channels. A disadvantage is that predators can likewise locate the sender. However, vibratory signals fade as rapidly as optical signals. When repeated at intervals or given in sequence, they convey extra meaning somewhat as the sequence of sounds does in human language.

Kinesthetic Signals

Kinesthetic or tactile signals are perhaps the most informative and important in developing and maintaining group social bonds and general group cohesion. Most often, they communicate a desire on the part of the sender to establish a peaceful bond with the receiver. Grooming, the major kinesthetic signal used by most primates, plays an important role in relaxing tensions, maintaining the equilibrium of the group, and reinforcing social bonds (Chapter 8). Tactile signals are prominent in mating behavior, mother–infant contacts, peer play groups, and in soliciting support in aggressive encounters.

Kinesthetic signals are clearly directional, since to be effective they require physical contact between the sender and the receiver. They cannot be broadcast; the possible exception is the huddling together of a group of monkeys as bonnet macaques often do (Rosenblum et al., 1964). The transient nature of these signals allows sequential tactile events to have specific meanings when used in combination.

ANALYSIS

In studies of free-ranging monkeys, we have gathered many social sequences that may indicate causal relationships between the signals sent and the resulting activity of the recipients. However, we have had to use laboratory experiments to test which elements of the messages actually transmit the information and stimulate the response recorded in the wild. These laboratory studies help us to analyze specific elements of the message as well as combinations of message elements; however, the field studies add to this an understanding of the context in which the signal occurred, which in turn adds to or changes the meaning of the message.

Signal context includes other accompanying signals, the internal and external condition of the animal itself, and the social context in which the signal is given. Sometimes signals stand alone, as when an intense warning bark cuts across other signals that are being emitted at the time; at such times, the social context is ignored as the animals flee to safety. But more often signals are emitted in series. Such signals modify each other, and the meaning of the whole message is more than the sum of its parts. For example, a play-face preceding signals that would otherwise be taken as a threat superimposes an additional meaning on the other signals, and the approached animal reacts with a play response rather than with the avoidance response that would otherwise occur. Signal context also includes the emission of parallel signals—two signals given at the same time. Macaques frequently emit a growl or screech during a facial threat gesture, and among bonnet macaques a growl indicates that the threatener is dominant, whereas a screech indicates the reverse. In any given social situation, the combination of the two signals means much more than the simple threat.

The emotional state of the animal emitting the signals also communicates additional information, just as an angry flush imparts additional meaning to what is being said by a human being. The physiological state of the animal is also communicated—for example, the estrous swelling of many cercopithecid monkeys.

The social context of these signals cannot be ignored, since the response of the receiver is affected by it. When a high-ranking animal threatens a low-ranking one with even a mild threat, the strong response is totally unlike the mild response given by an animal that is close in dominance rank to the animal that gave the signal. The presence of other conspecifics influences the response to a signal and therefore must have influenced the meaning for the receiver. Obviously, if the receiver of a threat signal can depend on the aid and support of nearby animals, his response to the signal will be affected accordingly.

EVOLUTION OF PRIMATE DISPLAYS

What was the developmental or evolutionary origin of primate displays? In an attempt to reconstruct this development, Andrew (1963, 1965) studied the function of primate muscles and compared the muscles of different forms. He concluded that displays often arise from initial movements of actions that are not carried to their conclusion and therefore are used as intention movements. In their simplest form, they are the incipient stages of an action: the beginning of a lunge for an attack, the opening of the mouth for grasping, the beginning of closing the glottis or of forcing the air out of the lungs. But if the signal so sent is adaptive, if it helps to maintain the social equilibrium that, in turn, helps the animal to survive, the actions themselves receive evolutionary emphasis through selection either by adding color to the face, tail, bare patches of skin, or the hair or by exaggerating the movement, for example, by enlarging the muscles and making the movements more pronounced than they need to be for the action itself. In animals that use their mouths to attack, the baring of the canines, the opening of the mouth, and a direct stare that orients the eyes on the intended victim are examples of the checked attack acting as a threat. These gestures are heightened among primates whose faces tend to be hairless, so that the relative positions of the eyes, nose, and mouth and their changing shapes are more obvious than in a fully furred face. Color differentiation adds emphasis, as when the white upper eyelids in a pink macaque face flash and thus intensify the signal.

In the estrous female rhesus macaque, sexual swelling, which is a signal of sexual receptivity, is another example of a signal that is emphasized by exaggeration. Among rhesus macaques and savannah baboons, this swelling is so pronounced that it can be seen for a great distance. Another macaque signal is the tail-raising of a dominant male. The one-inch stub of the Japanese macaque is in sharp contrast to the rhesus tail, which, though shorter than that of some other species (8-10 inches) is remarkably expanded in breadth by its long hair. When this bushy tail is raised to expose the bright red skin of the perineum and scrotum during the mating season, the message of dominance is obvious!

Andrew (1972) argues that gestural communication arises from intention movements in four categories of response: alert, protective, exertion, locomotion.

Alert Responses

Alert responses follow when the major sense organs collect information about the environment or a particular part of the environment. They arise when there is a stimulus contrast, or when what is perceived by the senses is different from what was expected. The animals know their home range well enough so that when any new object or animal supplies a different input, they respond by being alert; that is, each of the various senses receives enough information to compare it with past experience. The animal then makes a judgment to approach the object, eat it, run from it, or attack it. The alert response usually involves focusing the senses on the strange object and follows one of two general patterns.

The first concentrates on a single particular stimulus. With the eyebrows approximating, the obicularis oris constricting, and the ears pulled forward, the attention is focused on the object. These movements cut peripheral vision and focus hearing. In social communication, the alert reaction has been elaborated into a threat. A confident animal uses elements of the alert reaction to examine the social fellow. According to Andrew, this has been converted into a stare threat of varied intensity, depending on the number of elements involved. The white of the eyelid that shows in many species of macaques, baboons, and other Cercopithecidae is an example of a phenotypic character elaborated to emphasize the stare. Monkeys offer a mild threat by looking at the other monkey and a more intense threat by flashing the eyelid and by other facial and postural gestures.

The second general pattern of alert response is often used in protracted recognition comparison, when the individual is gathering enough information to check the identity of the stranger with a memory model from previous experience. The eyes are opened wide, the eyebrows raised, and the pupils dilated to let in more light and to cover a wider area. Most animals use this kind of alert response when the strange object is frightening or when they are not confident enough to focus in upon the object and ignore the surrounding stimuli. Apparently it is an attempt to expand the field of vision and to keep in view not only the frightening object but anything else related to it. Andrew argues that this response has been elaborated into subordination gestures. When threatened, primates look away but always keep their field of vision broad and the threatening individual in view on the periphery; thus they avoid giving any direct look that might be interpreted as returning the threat. During threat encounters, it is an advantage for the subordinate to have an expanded field of vision.

Protective Responses

Protective responses serve to protect the sense organs and other vulnerable structures from attack. The typically human protective reaction is to hunch over in a huddle and to cover the abdomen and face with the arms and hands while closing in the face with the facial musculature.

The first phase of the response is to close the eyes, partially or fully, lower the brows to protect the eyes, constrict the muscle that surrounds the eye (obicularis oris), and cover the ears. These responses are often accompanied by a check in respiration, a closing of the glottis with an expiration of the breath, and a retraction of the corners of the mouth. This latter reflex produces the fear grimace in many animals and, in a roundabout way, the smile in man. In some primitive forms, the teeth are somewhat exposed.

In an evolutionary elaboration of this signal, the muscles have shifted to pull the upper lip up and the lower lip down to retract the lips. Thus, the gesture becomes a very wide grimace that fully exposes the teeth. There is a marked difference, however, between the primitive threat gesture that virtually covers the teeth and the conciliation gesture that fully exposes the teeth. For example, there are different degrees of exposure among bonnet macaques. A mild subordination gesture involves lip-smacking, that is, opening and closing the lips slowly or rapidly with the corners of the mouth retracted. Extreme fear retracts not only the corners of the mouth but the lips in all directions to fully expose the teeth and gums. Such fear-grimacing is common among primates. A gesture intermediate between threat and conciliation is sometimes used by an animal uncertain about its dominance or used by two parties of a threat sequence that are nearly equal and hence uncertain about which will dominate. Bonnet macaques, for example, "jaw" (that is, open and close their mouths) by moving the mandible, not just the lips as in lip-smacking, and partially drawing the lips back. These two distinct gestures can be combined, although they evolved from alert responses on the one hand and protective responses on the other.

In the second phase of the protective response, the head or parts of the body are moved to remove foreign substances or objects. Repeated tongue protrusion removes such objects from the mouth, and repeated shakings, from the body. Because primates are oriented to visual facial signals, many species have elaborated tongue protrusion into a rhythmic sequence that has become part of the conciliation gesture.

Exertion and Locomotion Responses

Responses that involve and precede exertion have also been elaborated into specific signals: for example, crouching before a leap, tensing the musculature, increasing the blood flow to the muscles and initiating a series of physiological changes. In man, blushing is an example of exertion response; that is, the blood rushes to the surface of the skin as a cooling response. The opposite or warming response involves a huddle during which a subordinate animal, presumably cold with fear, retracts to conserve warmth while the dominant animal warms up and expands. Physiological changes include sexual swelling, which is seen in its extreme form in the estrous female rhesus.

In a similar manner locomotion and posture communicate social information. A dominant animal strides confidently, while a subordinate moves more hesitantly.

THE FUNCTIONS OF PRIMATE COMMUNICATION

Subhuman primate communication performs several major functions that help to coordinate the activities of the group. Social grouping is an adaptive mechanism that ensures the survival of the participating individuals by offering them protection from predators and an opportunity to reproduce. But if the group is to be more than an aggregation of independent and unorganized individuals, the members must coordinate their behavior for defense or successful flight from predators, for acquiring sufficient food, and for reproduction. Moreover, aggression among group members must be reduced and channeled in such a way as to ensure the least amount of social disruption.

WARNING

A warning system that enables all members of the society to escape safely is of primary concern to social groups living where the incidence of predation is high. For most primates, escape means flight to sanctuaries in the trees. In contrast, savannah baboons present a united front of big, aggressive males, behind which the weaker members of the society are protected. Among the terrestrial patas monkeys, a large male acts as a decoy while the other members conceal themselves in the tall grass.

The warning signal is so important that it must be easily recognizable and must cut through and drown out other incoming signals immediately. It must be a broadcast signal that can be received by all the members of the society. It must fade rapidly to minimize the possibility of the predator's locating the sender. Moreover, it is desirable that the warning signal be similar to that of other species, genera, and orders of possible victims so that the local community can be alerted by the warning call of any one species. On the other hand, the warning signal must be distinctive enough to bring about immediate and appropriate action in the conspecifics. A warning signal given by a crow, for example, may alert the monkeys in the area to the approach of danger without arousing them to full evasive action unless it is reinforced by a warning call from one of the monkeys.

This seemingly elaborate list of requirements can, surprisingly enough, be met by so simple a signal as the warning bark. First, it is so sharp and piercing that it cuts through other signals being sent at the time. Second, unlike a visual signal, which requires that receptors be directed toward the sender, it is broadcast and can be received from any direction. Moreover, it fades rapidly. Finally, the warning bark can vary in strength, so that distant danger can be brought to the attention of the group by a mild warning bark before last-ditch efforts to escape must be made. If the danger is sudden and imminent, that fact is communicated by a sharp, piercing bark that spells urgency. The precise tonal pitch that characterizes each of the various primate warning barks is species-specific and is easily recognizable by the members of the species, which are not likely to ignore it. In general, the various warning signals of primates are similar to those of birds and many other mammals; as a result, whole communities of animals can be alerted to take evasive action when necessary.

The vocal warning bark serves its purpose so well that for most species it needs no supplementary signals in other media unless the active response (for example, the sudden run for the trees) is regarded as a supporting signal.

TROOP MOVEMENT AND LOCATION

To function as a society, the individuals in a group must remain together. Such cohesiveness is maintained in part by signals about group movement and location. The seemingly simple process of keeping twenty monkeys in contact with each other is rendered complicated by the natural environment, which provides numerous avenues of concealment. The luxuriously foliated trees of the forest make it difficult to see more than a few feet in any direction. Even the open grassland of the savannah can swallow up a troop of baboons almost without a trace. If the grass is even two feet high, only individuals that are very close to each other can maintain visual contact. Signals are needed to apprise the individuals of the location and direction of movement of other group members.

A group of primates may separate temporarily into two or more divisions and may even spend several hours or stay overnight in different localities. A signal, usually a grunt or low-pitched growl, is required to bring the group together again.

Baboons on the savannah have a harder problem than arboreal species. Since the troop itself forms the main protective element for the individual, the baboon that loses contact with the troop is in grave danger, especially where trees are lacking. As a result, the moving baboon troop resorts to a steady stream of low grunts which carry for only a short distance. Thus the troop members move either within sight of each other or within sound of the grunts, whose low pitch not only serves to limit the troop's spread but also makes it difficult for predators to locate the troop.

Without vocalizations or other signals, the moving arboreal monkeys can audibly locate one another through the incidental shaking and rattling of tree limbs and leaves. Primatologists, for example, capitalize on this leaf-rattling propensity to locate moving troops of monkeys. The resting troop is quiet, however, and when movement begins, the group members must be recoordinated. A grunt or low growl is the usual signal to begin a troop movement. Begun by the first animals to move, it is taken up by the others as the whole group begins to shift almost in unison.

Reuniting separated parts of the group requires a different set of signals. The low grunts used to maintain contact within the troop on the savannah or to coordinate group movement from a resting position in the trees do not carry far enough for groups several hundred yards apart to hear each other. Macaques solve the problem by branch-shaking. One or more big males climb to the tops of tall trees and by forcefully shifting their weight back and forth several times cause the leaves to rattle loudly. This distinctive sound, which carries a considerable distance, helps the separated groups to rejoin.

The same signal is used by macaques to maintain spacing between groups. When two independent groups of monkeys approach each other, they resort to branch-shaking, which serves as a signal to move apart and thus to avoid conflict.

Same signal for group — ing & moving apart.

Since branch-shaking is of little use on a windy day and is satisfactory for only a hundred yards or so, several species of primates have elaborated more distinctive signals that carry over a longer distance. Gibbons, howler monkeys, and Nilagiri langurs, for example, have a special vocalization that is used primarily to space groups in relation to one another. Since location precedes spacing, the elaborate spacing signals of these species seem to be an extension of the simpler problem of location.

All three of these species (gibbon, howler, and langur) begin the day with a chorus of locator calls that are repeated by the other groups in the area. After the exchange of signals, the groups move out of their sleeping positions to feeding areas. As the day's activities proceed, howlers and langurs tend to avoid those areas from which the closest and most intense signals emanate. As a result, several groups feeding in the same area maintain a distance that eliminates contact and conflict between groups. The gibbon, however, reverses the procedure to reinforce the boundaries of the territory (see Chapter 3 for a detailed description). The signals are repeated intermittently throughout the day, and chance encounters between two groups of howlers usually provoke vocal battles that use the same calls.

Thus the focal problem of troop movement and location involves a series of related problems: unity within the group, avoidance of other groups, and coordination of group movement. Since all share the problem of location, the same signal can be used for different situations (for example, branch-shaking for both rejoining and group spacing).

SOCIAL FACILITATION

Communication to facilitate the social life of the group involves a multitude of activities. Here it is limited to those signals that smooth the way for any social activity and reduce the possibility of aggression. One of the most important social problems is the control of intragroup aggression; its magnitude varies with the species. Baboons, for example, which need highly aggressive males to protect the troop on the open savannah, are in a more difficult situation than exclusively tree-dwelling species, which usually resort to flight rather than fight. But any social group encounters situations in which some commodity is desired by more animals than the limited amount can supply. A single estrous female cannot simultaneously copulate with two males. Even if she copulates with them in succession, the question of who goes first must be settled. Dominance ranking supplies the answer, but it must be communicated.

Four levels of communication are involved in social facilitation. On the first level, facilitation signals are transmitted through the tactile medium and are generally neutral in terms of dominance. Grooming between group members of all ages, play that involves rough-and-tumble wrestling, the clinging of an infant, and huddling in sleeping and resting positions are all parts of the tactile communication at this level that soothes, relaxes, and reinforces social bonds. These activities are discussed in detail in Chapters 8, 9, and 10.

On the second level, facilitation signals are transmitted through the vocal–auditory and visual channels to reinforce the established set of social relationships without conflict or marked tension. The stronger the ranking system in the social group, the more common these signals will be in everyday life. Most social interactions take place within a framework made explicit by these signals.

As on the second level, facilitation signals on the third level are made through the vocal–auditory and visual channels. However, they reestablish social equilibrium disrupted during conflict. They communicate threat and conciliation as the individuals work out the differences among themselves and settle back into a regular routine or establish new relationships.

Fourth-level facilitation signals involve physical aggression and are used only when the conflict cannot be resolved by threat and conciliation. As on the first level, they are transmitted tactilely; thus they represent a full-circle return to a medium that is unequivocal and difficult to misinterpret. Both visual and vocal signals can be sent to numerous individuals, some of which are not intended to be receivers. But a tactile signal is much more difficult to misdirect, whether it is an infant's nuzzling its mother's breast to communicate its desire to nurse (first-level) or a hamadryas male biting the neck of one of his females to force her to follow him again (fourth-level).

First-level facilitation signals are treated later in Chapters 8, 9, and 10. Fourth-level signals are simple and direct and will not be elaborated upon here. For the most part they involve various kinds of bodily attacks. Second and third levels of facilitation communication will be treated briefly here.

Second-Level Facilitation Signals

When approaching each other, two bonnet macaques of nearly equal rank, particularly if they are males, are likely to give second-level facilitation signals to reinforce their relative positions. If the more dominant of the two approaches confidently, moving forward in a relaxed manner without looking away, the subordinate signals his acceptance of the relative ranking by turning or looking away. Thus we have paired opposed signals (Figure 7). Assuming that the rank of both animals is about equal, a signal from either of the pair will normally elicit the opposite signal in the other. If the situation is more intense, as immediately after some social disruption, more intense second level facilitation signals are exchanged. Figure 8 ranks the intensity of second-level facilitation signals among bonnet macaques; the higher numbers represent higher intensity in both dominant and subordinate signals.

But when two animals are not close in rank, the differences between them must be taken into account. For example, a more dominant animal adds his distance from the subordinate animal to his signal, so he need not give so intense a signal to a low-ranking animal as he does to receive the same response from a higher-ranking subordinate. An alpha male bonnet macaque reinforces his relationship with a low-ranking subadult male by the confident approach and, because of the rank difference between them, may receive a present and grimace instead of a simple look away.

Dominant signal	1	Confident approach
Subordinate signal	1	Turn or look away

FIGURE 7 Paired, opposed second-level facilitation signals are mild and are used in situations that require little reinforcement.

	4	Full mount
Dominant	3	Symbolic mount (hands placed on other's hips)
signals	2	Approach with tail raised
	1	Confident approach

	1	Turn or look away
Subordinate	2	Move to avoid
signals	3	Present
	4	Present and grimace

FIGURE 8 Paired, opposed second-level facilitation signals among bonnet macaques. Higher numbers indicate a greater intensity in terms of social meaning.

Third-Level Facilitation Signals

Second-level signals do not always ensure that conflict will be avoided even though they strongly reduce the possibility of such conflict. Many situations arise that, either because the dominance of two individuals is changing or because the presence of another individual subtly alters the social setting, necessitate an overt and forceful indication of dominance. Here again the dominant signals are balanced and paired with opposed subordinate or conciliatory gestures that, when used by individuals of nearly equivalent rank, can restore the social situation to equilibrium. Figure 9 shows bonnet macaque signals on the third level of facilitation. Again, those of equal degree of intensity are paired opposites. On both sides, the signals are continuous, but dominance signals lend themselves more clearly to discrete divisions than the subordinate.

Among bonnet macaques, third-level signals are used with vocal modifiers. The continuum of vocalizations runs from a high-pitched squeal through screeching and into a series of growls. The squealing represents extreme fear and is used almost exclusively with fear gestures (the various grimaces from 3 to 6 in Figure 9). Screeching is an intermediate modifier that when used with subordinate signals indicates fear-subordination; when used with dominant threat signals, it indicates both fear-subordination and threat. If two animals are threatening each other, the dominant one growls and the subordinate screeches. If their relative dominance changes, their vocalizations also change, even though their gestures

	6	Attack
	5	Lunge
Dominant	4	Open-mouth threat
signals	3	Eyelid threat
	2	Stare
	1	Look

	1	Lip-smack slowly
	2	Lip-smack rapidly
Subordinate	3	Grimace
signals	4	Grimace widely
	5	Grimace and present
	6	Run off, grimace, and tail-whip

FIGURE 9 Third-level facilitation signals of bonnet macaques (*Macaca radiata*).

PLATE 14   Japanese macaque (*Macaca fuscata*) giving an open-mouthed threat, one of the third-level facilitation signals.

remain threats. The paired opposed signals come into play as the sequence of threats resolves into a stable social interaction.

When the dominance of the two individuals differs greatly, third-level signals have a factor added or subtracted, depending on the rank of the individual transmitting the signal. For example, on a dominance scale of 0-5, a total of 11 can be used to indicate a highly dominant animal attacking: 6 (see Figure 10) for the signal intensity of the attack and 5 for the individual's dominance. (The assignment of values is arbitrary with currently available evidence.) If the conflict is to be resolved immediately, an equal subordination score must be attained by the attacked individual. Only another animal with a rank score of 5 could resolve the situation by using the subordinate signal of running off, grimacing, and tail-whipping (a score of $6 + 5 = 11$). Theoretically, such an animal could return immediately to close social relations with the attacker. Since most animals attacked by a high-ranking animal cannot attain such a subordination score, they must run off and avoid the attacker for some time before reestablishing social relations.

PLATE 15   Pig-tailed macaque (*Macaca nemestrina*) giving a jaw-thrust gesture that is often used in sexual solicitation by males. (*Courtesy of the Oregon Regional Primate Research Center; taken by Harry Wohlsein.*)

As another example, if an animal with a rank of 5 stare–threatens another with a rank of 2, the subordinate this time can resolve the situation by grimacing and presenting. Thus rank 5 plus stare–threat 2 equals 7; rank 2 plus grimace and present 5 equals 7, and the social equilibrium is reestablished.

Several individuals can coordinate their activities and join to challenge another individual or coalition. Thus two lower-ranking animals can successfully dominate a higher-ranking animal who has no support. This can be elucidated by the following equation: (individual dominance + signal intensity) + (individual dominance + signal intensity) = individual dominance + signal intensity). If all signals are threats, the side of the equation that is higher wins the encounter.

Fourth-Level Facilitation Signals

Third-level facilitation signals give to the society a flexibility to adjust social disequilibrium before physical aggression occurs. Second-level facilitation signals tend to maintain the status quo, and first-level signals are positive social bonds. When all other signals have failed, the last resort is physical aggression. But since the fourth-level facilitation signal

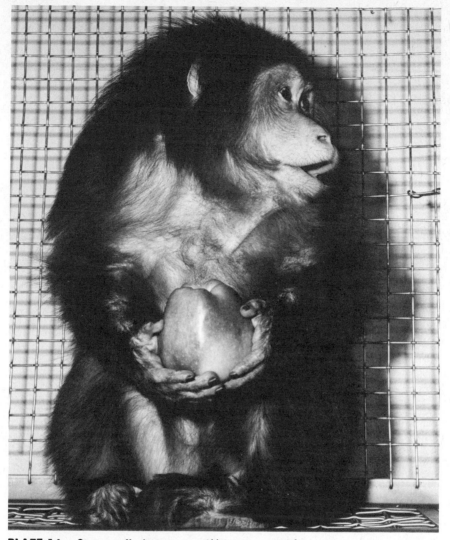

PLATE 16   Stump-tailed macaque (*Macaca arctoides*) giving an open-mouth threat. (*Courtesy of the Oregon Regional Primate Research Center; taken by Harry Wohlsein.*)

is a tactile one, aggression, it reestablishes physical contact and usually results in turning the aggressive contact into a new close-communication sequence that releases the aggressive elements and restores the social situation to equilibrium.

SOCIAL COMMUNICATION

It is difficult to conceive of any society without communication, so difficult, in fact, that society can be seen as a matrix of communication networks. Here space is too limited to permit a detailed comparison of systems of communication found among subhuman primates. Hence they are treated where aspects of social structure are discussed in the later parts of this book. Tactile communication during play, grooming, and the mother–infant relationship has been specially emphasized because of its importance in establishing the social bonds that form the substratum of nonhuman primate society.

Chapter 7
Social
Organization

Assemblages of animals can be either chance aggressions of individuals that may never again assemble in the same configuration or an integrated social organization with some degree of permanence. The social organization itself can be temporary or permanent, seasonal or annual. It can be the focus of most of the individual's activities or a passing phase during which he restricts his social behavior to some specific activity like reproduction, after which he resumes his independent status to pursue other goals like food and survival.

Eisenberg (1966) makes the following distinction between chance aggregations and ordered societies.

An integrated social organization must be distinguished from a random aggregation by means of several criteria. These include: (1) a complex system of communication, (2) a division of labor based on specialization, (3) cohesion, or a tendency of the members to remain together, (4) a permanence of individual composition, (5) a tendency to be impermeable to conspecifics that are not members of the group. Integrated social groupings may be formed in two ways, either as an outgrowth of a family group or as a result of outside recruitment of unrelated individuals into a cohesive unit.

Virtually all primates live in integrated social organizations and meet the behavioral criteria argued for by Eisenberg. So far as we know, those that do not are restricted to isolated males of species whose modal pattern is integrated social organization.

Almost everything a social animal does serves as communication with its fellow group members. The complex systems of primate communication described briefly in Chapter 6 tell much about their environment, both social and physical.

The simplest form of the division of labor separates the roles of the sexes in reproduction; beyond that there is specialization within the sexual roles. In some primate societies, the division of labor is based on age differences. For example, in larger baboon and macaque groups with numerous males, young males often serve as front or rear guard for the group as it moves. The fully adult, dominant males, which are the group leaders, remain in a more central position, ready to move in any direction toward threatening danger or perhaps to lag behind as the rest of the troop flees danger. No real economic division of labor exists among primates other than man since generally each individual, once it outgrows its dependence upon the mother as a source of food, is an economic unit.

The cohesion that characterizes integrated societies and binds the members of a group is sufficient to hold the females of most species within the group throughout their lives, up to twenty years or more. Although most males also remain in the group, in some species the males shift from group to group rather frequently, and in others the males often become isolated. But a high degree of tactile contact early in life and, for many species, persistent tactile contact through grooming, play, and sex tend to establish and reinforce the cohesive social bonds.

Since, in most cases, the cohesive tendencies of primates are lifelong, the social group consists primarily of individuals that are born into the group rather than of outside recruits. Occasionally (typically in some species of macaques) members are harried out of a group, but the group remains integrated; only rarely do upheavals radically disrupt the social contacts among individuals.

Finally, since membership derives primarily from birth, outsiders often find it difficult to become members of a primate society. The primate social group is fairly impenetrable. Even chimpanzees, whose social groups were originally thought to have totally permeable boundaries, actually have a group structure that is similar to that of many other primate species. The chimpanzee social organization is simply more flexible within the group; and subgroupings, which change membership from day to day, are observed more often than the whole group. In fact, for effective social organization, the group size must be limited. Each individual must be able to recognize and predict the behavior of every other member of the society. If there are too many mistakes, the whole social organization breaks down. The unlimited exchange of individuals from group to group would destroy the predictability of behavior.

PRINCIPLES OF SUBGROUPING IN PRIMATE SOCIETY

Several factors are important in the formation of primate societies, some of which are basic to the formation of societies in general and are found among other mammals and vertebrates as well. The mother–infant bond is certainly an important and primary social bond for mammals. In the wider category of vertebrates, the reproductive relationship between male and female forms a primary social tie. But the basic sexual and mother–infant relationships can be widened by other sets of relationships and reinforced by other bonds to form an elaborate social organization of both sexes and all ages.

For vertebrates in general and mammals in particular, three strong social bonds help to build a sturdy and viable society. The sexual relationship, when elaborated to form a permanent-pair bond, often results in societies consisting of a single adult male and a single adult female. This is a common form of society in passerine birds, in which the males as well as the females care for and feed the young from the time they hatch. Although such pair-bonding is found among primates (gibbons, for example), it is not the most common form of social organizing force for primates or mammals in general.

The second major factor in primate society, the mother–infant bond, is central in mammalian social relations. The specialized infant feeding done by the mother is not usually duplicated by any male activity; therefore, in many mammalian societies, the male is less important to the social group. The mother–infant relationship can be forged into a strong, permanent bond, and a group that is larger than the single dyad can be developed. I call this unit the mother-lineage since it is based on the relationship of the infant to the mother and forms a primarily lineal association between the individuals. Since it is exceedingly widespread among primates, the mother–lineage is the primary social organizing force, and the differences in social organization are, in most cases, modifications of the basic mother–lineage.

The third major factor in primate society is the social bond between adult males. This is the adult male association that plays an important role in the societies of many ground- and semi-ground-dwelling primates. It is the least common of the three since it does not play any direct role in procreation and in the support of the young in their period of total dependency.

THE MOTHER-LINEAGE

The mother-lineage is a group of conspecifics, all of which are related to a common ancestor through the female parent and have developed their associations through the mother–infant bond. Descriptively, the mother–lineage contains an old female and her mature daughters with their offspring, both infants and juveniles, of both sexes. It can also include associated subadult sons who usually, as they grow older, spend

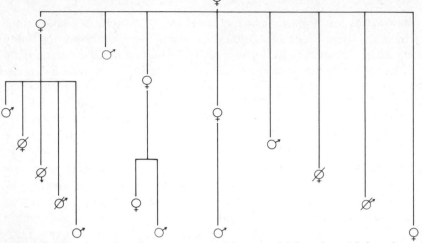

FIGURE 10   A mother-lineage centered in one old female, with her descendants over a period of eight years. (⌀ or ⌀ = deceased)

more and more time away from the mother-lineage in other activity. The mother-lineage is strongly female-oriented and can operate as a self-sufficient social unit provided there are occasional contacts with adult males to provide fertilization.

Such a group can contain three to four living generations and is usually centered in the oldest living female of the lineage. Her death is probably the most common cause of fission. Since the mother-lineage coheres because of the mother–offspring relationship, the death of the oldest female severs the primary bond holding the next younger generation of females together. The lineage then tends to break into separate lineages, since this next younger generation of females has stronger bonds with its own offspring than with its siblings.

Nevertheless, although the mother–infant bond is basic, lateral relationships are formed between siblings that are brought into close contact through strong social contacts with their mother. In their youth, siblings and, as the mother-lineage grows, the offspring of the other adult females in the mother-lineage are in more regular contact with one another than with other individuals from other mother-lineages unless there are positive mechanisms drawing the young into extra-lineage contacts. Since this sibling contact occurs early in life, it is more likely to form a solid bond among individuals than associations formed later in life (assuming that the latter bonds are not formed at the expense of the mother–infant bonds).

Thus, as the mother-lineage begins to expand with new offspring, with the maturation of the daughters of the original female, and with the production of their own offspring, a network of social bonds is formed based primarily on the relation between mother and infant, secondarily on the close social relationship of all the members of the group.

This creates a powerful autonomous social unit. In fact, several mammalian species rely on this as the primary social bond that has the permanence to hold the group together throughout the year. In the red deer (Darling, 1937), the mother-lineage moves together as a group throughout the year; only during the rutting season do the males join the group. Even then, they do not act as an integral part of the group; instead, in times of danger, they usually go their own way, leaving the females to fend for themselves. In areas where small foraging parties are primarily adaptive, the mother-lineage forms the only stable grouping. The vast herds of ungulates in Africa and the great plains of the United States before they were decimated probably consisted of clusters of such mother-lineages bound together by some other forces. Such bonds were either permanent or temporary, affording greater stability on the one hand and greater flexibility on the other.

Permeability and Impermeability

One feature distinguishes the mother-lineage from other social forces: It is, strictly speaking, an impermeable social group. Except for rare adoptions, which are most often confined to members that are already a part of the mother-lineage, an individual is born into the mother-lineage. The very nature of the bond forming the mother-lineage precludes outsiders. In contrast, the adult male association is drawn from members of various mother-lineages and is formed by their interaction outside the mother-lineage. Actually, members of other groups can also be incorporated in the adult male association of a particular troop of monkeys. Play groups and grooming relationships sometimes include members added any time from any source.

Rank and the Mother-Lineage

In the larger social groups with numerous males, social tensions are usually eased by ranking systems that predetermine priorities in the acquisition of limited resources: food, sexual partners, desirable resting places. The mother-lineage plays an important role in determining an individual primate's dominance. An infant is born without any basic rank of its own, but its mother's rank gives it social status since she will support it against even higher-ranking members of the group. Kawai (1958a) says, "One of the most important factors [in establishing the individual's basic rank] may be the effect of the dependence of the infant period which is influenced by kinship." Studies on Japanese and rhesus macaques show that, in these species at least, the position of the mother and of her close relatives has a very strong effect on the development of the offspring's rank, particularly of those near the core of the troop.

The influence of the mother extends to peer relationships since, as Sade (1967) has shown, rhesus infants ". . . defeat their age peers whose mothers rank below their own and are defeated by their age peers whose mothers rank above their own." Thus strong mother–offspring bonds, embedded in a matrix of social relationships such as the mother-lineage, enable the growing primate to interact with outside members of the society with the secure backing of several animals.

Most studies of rank have centered on the males, since their dominance is clearer in macaque and baboon societies. But in societies where the mother-lineage is strong, the females are clearly in a position to strongly influence the group. Discussing the formation and development of the mother-lineage among Japanese macaques, Imanishi (1957) notes that because of their close identification with their mothers, the daughters of high-ranking females tend to remain in high-ranking positions and to hold them on their own when the mothers die. Imanishi believes that the females definitely influence the choice of leading male in the group and that the central females, high-ranking and close to the adult male association, determine whether a male, who may be high-ranking on his own, may enter a group or not. Thus a male's ties with his mother may give him the dependent rank to develop a high basic rank of his own, and his early bonds with female siblings may determine whether he is allowed to enter into the center of the troop to exert his rank.

Japanese Macaques, Rhesus Macaques, and Chimpanzees

Only three species of primates have been carefully and closely studied over a sufficiently long period for the structure of the mother-lineage to be examined: the rhesus macaques of Cayo Santiago (Koford, 1963; Sade, 1967; Kaufman, 1967), the Japanese macaques (Imanishi, 1957; Yamada, 1963; Kawai, 1958), and the chimpanzees (Van Lawick-Goodall, 1967). Among chimpanzees, the mother-lineage is the most stable and cohesive subgroup, but it does not persist as permanently as that of some of the rhesus and Japanese macaques. The chimpanzee mother-lineage consists of the mother, her older adolescent offspring (male and female), and her recent young. They are usually found together, move together, and act as a stable social unit. But as they reach maturity, both the female and male offspring spend more time in other groups, and the stable mother-lineage does not really include more than one adult, the original mother.

Among Japanese and rhesus macaques, the mother-lineage is not the first and obviously recognizable social group. But closer examination reveals that for many of the individuals in the society it forms the nucleus of social relationships. A mother associates most closely with her new-born infant, which, as it grows and reaches adulthood, continues to have a higher percentage of its social relationships with its mother and the siblings with which it associated while it was growing up. The mother-lineage has such strength in macaque society that it is capable of over-riding other social relationships when a troop divides. When the large troop of Japanese macaques at Takasakiyama, for example, reached a size that made it impossible for all the individuals to know and react socially with one another, the troop split. At first, many of the older females, who had occupied the center of the troop near the dominant males, remained with these males; but those whose daughters had departed into the new troop (led by peripheral subleaders of the original troop) soon joined their daughters, thus maintaining the integrity of the lineage at the expense of their relationships with the dominant males. This indicates that the mother-lineage is a strong and powerful social unit and that the bonds developed between mother and infant are exceedingly important in the socialization of the primate.

Means of Binding Mother-Lineages Together

Some primate societies, as will be shown shortly, function with one adult male, and there is no real difficulty in developing mechanisms for attaching a male to a mother-lineage (see Pair-bonding below). But other forms of primate society, particularly those of species that spend some time on the ground, must rely on a large number of individuals in the social group, particularly males, for protection against predators. Because of this, the mother-lineage cannot be the sole basis of primate society. If a viable social unit is to be formed, the mother-lineage must often be combined with other mother-lineages. The individual intolerance for those from other mother-lineages must somehow be overcome so that bonds cutting horizontally across the vertical bonds of the mother-lineages can be formed. Such bonds are numerous, and they form a matrix that can be exploited in a variety of ways. Play, grooming, and sex, which are discussed at length in later chapters, are factors that build social bonds.

Other factors are discernible but not clearly described as yet. For example, the mangabey males (Chalmers, 1968b) spend considerable time caring for, holding, and grooming their infants. The close contact between the male and the infant forms a social bond outside the mother–infant relationship and links them into a larger and more complex society. Unlike male red deer, which are relegated to the status of occasional visitors, mangabey males are integrated into the society.

The stage of infant rejection by the rhesus mother is a factor in producing ties outside the mother–infant dyad. At that point of life, the young juvenile must build social contacts with other members of the group; although many of these contacts will continue to be with members of the mother-lineage (older and younger siblings and maternal cousins as well as the mother), the opportunity for contact with other animals outside the mother-lineage is important. It is as a juvenile in early play groups that the macaque and baboon form alliances and relationships with other members of the troop that sometimes persist as long as the individual lives. Yamada (1963), who has demonstrated play relationships in Japanese macaque juvenile peer groups (see the chapter on play), indicates that the young primate, despite a tendency to maintain a number of relationships within the mother-lineage, is not yet so rigidly related to the one group that it cannot form relationships outside it.

PAIR-BONDING

In spite of the flexibility of the mother-lineage, it is incompatible with strict pair-bonding as a form of social organization; although pair-bonding is not widespread among primates, it occurs often enough to warrant some consideration.

Pair-bonding during adulthood forms a new social bond between a male and a female that supersedes previous infant and juvenile associations. Thus societies that rely on pair-bonding as the pivotal force in social organization must have mechanisms for breaking the bonds between mother and offspring or at least for superseding them to such an extent that the mother–infant relationship is reduced to insignificance for the adults in the society. For such primate societies, the mechanisms whereby the mother–infant bond is broken are as important as those that form the pair bond. Neither mechanism has been studied satisfactorily, but the gibbon family is the best example of the pair bond at present.

Gibbon society is based on the permanent association of one adult male and one adult female. Two attractions that are potential binding forces must be broken if the pair is to form a permanent relationship without the complications of added adult members to the group: the mother–infant bond and the possible sexual bond between the growing subadults and their parents.

Gibbon males and females show little sexual dimorphism in body size or canine teeth. Female canine teeth are almost the same size as those of the male. This means that both the male and female are capable of exerting equal physical force in the society. In fact, Ellefson (1967) noted examples of both male and female gibbons exerting dominance over the mate of the opposite sex.

Offspring approaching adulthood are slowly forced out of the natal group. The process is probably initiated by maternal rejection, such as Harlow demonstrated for rhesus macaques, and is continued without the formation of other bonds to counteract the rejection. Whatever the mechanism, the subadults leave and apparently establish new groups with other rejected offspring. Pair formation has not yet been observed in the wild, so we do not know exactly what form that process takes.

One characteristic of a pair-bond society is that it is not self-sustaining. For each generation, as the old parents die off, a new society must be formed. Although mechanisms for continuing a group with the offspring of the parents may exist, this method of forming a new pair bond has not yet been demonstrated and probably is not common.

The pair must be embedded in a matrix of wider social relationships even if those relationships are only intermittent and aggressive. There must be other nearby groups rejecting offspring if the young adults are to come together and form a new pair. The majority of new pair-bond groups are probably formed by offspring that have been rejected by their natal groups.

The mother-lineage is thus effectively eliminated since the mother–infant bond is maintained only for a relatively short period of time. No female offspring may grow to adulthood and raise her own offspring while still directly associated with her own mother. The pair bond is not a satisfactory base for the establishment or development of larger, more elaborate social groups. For example, in the form manifested by gibbons and many species of captive tree shrews, the maintenance of the pair bond depends on the destruction of other social bonds, and this restricts the opportunities to form bonds between pairs and build a more complex social organization. By contrast, human pair-bonding relies on the maintenance of the other ties rather than on the destruction of competing ties.

ADULT MALE ASSOCIATIONS

An adult male association is a social subgroup whose members regularly support one another in their various roles as high-ranking leaders and the agents of social control. In large troops, such an association usually includes only a part of the adult males.

Being the focus for social relations, the adult male association is the ultimate authority in the group. Where it is strong (particularly among baboons and macaques), it is the overriding social influence; as the focus for the activities of other subgroups and individuals, it enables the society to become far more complex than would be possible with simply the pair bond or a one-male group. It is found chiefly among ground-dwellers like the large and elaborate troops of baboons and macaques and the semi-ground-dwellers like *Cercopithecus aethiops* and *Cercocebus torquatus*. Characterized by marked sexual dimorphism and composed of large, aggressive males, it functions primarily to protect the troop against predators.

The adult male association sharply defines the roles played by its members. The younger males approaching adulthood do not yet belong to it. Peripheral to the troop in many cases, they serve as an outer protective ring for the other members. In this role, they often display the same kind of bravado that characterizes the human adolescent in the presence of danger. Here, among wild primates, this bravado is highly functional, since it serves to draw attention away from the females and the young, thus giving them time to escape and giving time to congregate and mob the predator if they are capable of attacking (baboons against leopards and cheetahs, for example) or to form a protective rear guard as all group members make for safety.

Some males never become members of the adult male association, because they lack either the strength to compete physically or the personality to form the necessary social bond with other males, or perhaps both. In any event, their influence on the society is far less than that of the males who are associated. In many macaque societies, these males become isolated or at least peripheral to the group; however, some may develop the proper characteristics later and become central males and part of the adult male association.

Adult male associations are sometimes remarkably permanent. The six males that formed the leader group in the Takasakiyama troop of Japanese macaques in 1949 retained their leadership for twenty years, playing the leading roles in the society even when troop splits were necessitated by unmanageable size. Only the death of most members caused a change.

The members of an adult male association are ranked according to dominance, but because of their close social bond, there is little conflict among them. Mild threat gestures are usually sufficient to effect the adjustments due to rank. Nonmember males generally experience much more violent interactions with the associated males. But the differences in operation among troops of the same species show that the accommodation between associated males is not always identical or perfect.

Let us, for example, compare three troops of bonnet macaques. One troop living near the railway station in Coonoor, Tamilnadu, apparently had an adult male association that worked very smoothly. In a troop of about fifty animals, three high-ranking males were closely associated, often sitting in actual contact with one another. Once, when fed an orange, the second-ranking male picked it up and began to peel it. When the first-ranking male reached for it and sniffed his face, the second male handed it to the first-ranking male without leaving or showing extreme signs of fear, only using a mild lip-smack to show his subordination. All three sat together, in tactile contact, as the first-ranking male ate the orange.

In the Somanathapur troop, somewhat larger than the Coonoor troop (fifty-eight at the time) but similar in overall composition, the adult male association was in flux and the beta male could not or would not pick up offered food within twelve to fifteen feet of the alpha male. In fact, on one occasion, the alpha female (which had the constant support of the alpha male until he was driven from his position) picked up several offerings of bananas right in front of the beta male while the alpha male was in sight but on a branch fifteen to twenty feet away.

A third bonnet macaque group, which lived along the road between Bangalore and Mysore, apparently lacked any adult male association. This group had no subadult males or older juvenile males; only one adult male was present with twelve females and their young offspring. This male was so aggressive that when I fed them groundnuts in order to count them, he permitted no other members within twenty feet of the area where the nuts were spread. All other troops that I counted in this manner except one congregated on the ground and ate the groundnuts together. I concluded that this male was so aggressive and so unable to form social bonds with other males that he permitted no other males to remain in the troop.

The adult male association sometimes consists of males that are not the highest-ranking males but whose cooperation enables them to maintain control against high-ranking males that have no support. The Somanathapur troop of bonnet macaques, for example, in the summer of 1963 had two high-ranking males, each of which outranked any one of the four males in the adult male association. But these two males did not have the support of any other members of the troop nor did they support each other. When each came individually into the center of the troop, he was regularly forced out to the periphery if his presence caused any threats in the troop. The four associated males threatened cooperatively, using the vocalization of subordinates threatening a dominant animal while they forced the single high-ranking male to the periphery.

Thus the primary characteristic of the adult male association is a strong social bond between males whereby they consistently cooperate to maintain social control of the group. The nature of the social bond is not known but a few hypotheses have been formulated.

According to Imanishi (1957), the infant Japanese macaque easily identifies with its mother and learns many aspects of social behavior by imitating her and being rewarded by pleasant social interaction or the discovery of good things to eat. Although the infant monkey finds it more difficult to identify with a specific adult male, he must do so if he is to learn the proper behavior patterns and is to be socialized in male behavior. Imanishi also feels that for the young male monkey socialization involves absorbing the personality of one of the leader males if he is to develop his superego. His primary rewards for imitating the leader are self-satisfaction and a feeling of superiority over the other animals in the group, both of which are extremely difficult for the primatologist to demonstrate. If such an identification is made, the young male may eventually work into the adult male association without too much conflict. The position of his mother also helps, since a high-ranking female keeps her offspring in fairly close contact with members of the association that are excellent prospects for her young to identify with.

Chance (1961) and Chance and Mead (1953) describe the process of equilibration whereby monkeys coming into close contact maintain this contact through proper communicatory signals. Instead of resorting to flight when in contact with a higher-ranking animal, the younger male achieves social equilibrium with the higher-ranking animal by means of signals.

Identification and equilibration are probably only parts of a complex system that forms a strong bond between some adult males. Both are of importance for the young male easing into an established association with older generations of high-ranking males. It is possible that new adult male associations are formed in the play group as young males establish firm bonds with playmates. Allied young males can then displace the older males from the center of the troop. Many more factors must be involved in the formation of the special bond that unites the members of an adult male association.

FORMS OF PRIMATE SOCIETY

The forms of primate social organization vary widely. Part of this diversity is probably determined by the inherited range of species behavior, that is, the patterns that are possible with the inherited capabilities of that species. But the variation results largely from the ecological adaptations that must be made, tempered by the individual characteristics of the group members. at a particular time and the past history of their social interaction. Thus change in environment often brings a change in social organization, an adaptation to new conditions. These adaptive relationships are most clearly seen and are probably strongest in actual fact where environmental pressure is great: intense predator pressure (for example, baboons on the savannah with leopards, cheetahs, and lions), limited food supply (for example, hamadryas baboons living in near-desert conditions), competition with other forms for the same source of food (for example, vervets competing with elephants for the same fever trees). Where the species is not subject to heavy predation, the range of social organization within the same basic ecology is greater.

Primate social organization can be classed under four major headings: solitary individuals, mated pairs, isolated one-male groups, and troops (including those with strong adult male associations, those with one-male subgroups, and those with permeable subgroups).

SOLITARY INDIVIDUALS

The solitary life is only relatively solitary for sexually reproducing mammals. The solitary individual lives most of its adult life alone with only occasional territorial contacts with neighbors, an occasional contact with members of the opposite sex for mating, and, for females, the social contact of mother and infant. Even this minimum of social contacts can be rather extensive for the female, which spends weeks or months in close association with her mother while she is young and much of her adult life with her own offspring. The male can lead a more solitary life than the female: His social contacts can be limited to courtship and copulation.

No primates have been conclusively shown to lead solitary lives. The aye-aye (*Daubentonia madagascariensis*) and perhaps some of the other nocturnal prosimians do so, but without full confirmation of careful field studies it is unwarranted to assume that they are solitary. It is very difficult in broad daylight to see all or even most of a group of tree-living monkeys. It is much more difficult at night. It is quite possible that a pair of nocturnal prosimians forage separately or that their association is seasonal but nonetheless permanent.

In several species of macaques (for example, *M. fuscata* and *M. mulatta*) solitary individuals form a sort of secondary social organization. A number of growing males expelled from the social group become isolated from the troop for a period of years, even, in rare cases, throughout their lives. But some of them often form small unisexual groupings.

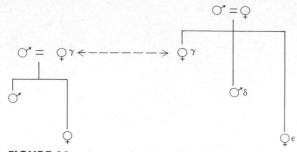

FIGURE 11 A mated pair with offspring. ♀γ is ejected from her parental group and forms a new mated pair with an outside male.

MATED PAIRS

Mated pairs, the simplest form of permanent social organization among nonhuman primates, involve a permanent bond between a single male and a single female. The social group then does not persist beyond the death of one of its members. The offspring of the pair remain with them only until they approach adulthood and then leave the group or are ejected from it by their parents. They then form new pairs with other recently rejected offspring and if space is available establish a new territory. Because the social bond holding the group together is the single exclusive bond between the male and the female, the size of the group is strictly limited and automatically excludes more adults. The social interaction between groups is usually antagonistic and takes the form of territorial disputes. Both gibbons (Old World apes) and *Callicebus* (New World monkeys) have such intergroup relations between mated pairs.

ISOLATED ONE-MALE GROUPS

More complex social organizations are possible if more than one female is included in the society. The male-female bond cannot then be an exclusive bond at the expense of all other adult social bonds. If the group's male complement is restricted by antagonisms among males, the tendency to drive out young males will persist; but the bond between mother and daughter can be exploited to overcome any tendency to reject the female offspring. For example, Hanuman langur infants are closely associated with the adult females, who regulate their play behavior (Jay, 1963; DeVore, 1963); the males are simply not involved. Thus whereas a strong bond develops between mother and offspring, the male–infant bond is weak. Denied any close contact with their group's male complement, young males cannot as effectively identify with the male element as young male Japanese macaques can.

Isolated one-male groups are primarily associations of females to which a male is added. The bonds of the mother-lineage are strong enough to prevent a pair bond from forming and from eliminating other adults.

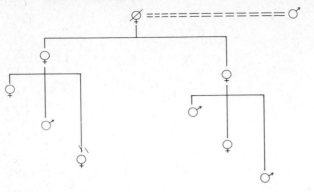

FIGURE 12   One-male group of Hanuman langurs hypothetically made up of a mother-lineage (with the original female deceased) and an associated single male.

Unfortunately, studies of species that have this social organization have not been carried on long enough for the kinship relations to be known. Therefore whether all the females are related or whether unrelated females make up the group is not known; both are distinct possibilities.

Two examples of the isolated one-male groups, the patas monkeys and the Hanuman langurs of Dharwar, show rather different internal relations; only the superficial structure is similar.

The Dharwar langurs have groups that contain only one adult male, several adult females, and their young (Sugiyama, 1964; Sugiyama, Yoshiba, and Parthasarathy, 1965). Female infants are passed among the adult females at birth and thus are socialized into a strong social group from the beginning. Moreover, they maintain these close social relations with the adult and subadult females. The male infant is socialized in much the same way when he is young but must be ejected from the groups by the dominant male as he approaches adulthood. The adult male joins a stable set of social relationships, and so, although he may dominate and control the group, in a sense he is an outsider. In an area with several isolated one-male groups and one all-male group, Sugiyama et al. found that a male from the latter group attacked, dislodged, and replaced the male leader from one of the one-male groups. The successful aggressor was, moreover, accepted by the females, and thus the continuity of the group was ultimately maintained by the females.

Among the patas monkeys (Hall, 1967), the solidarity of the female group is apparently not so great. The female patas monkey does not pass her infant among the adult females; it is therefore not socialized by close contact with other adult patas monkeys. Nevertheless, the females form a strong ranking system among themselves, and Hall suggests that, ". . . it may follow that the initiative in keeping a male within the group or allowing one to be ousted from the group comes through the female hierarchy." Thus, unlike the example cited by Sugiyama, the acceptance or rejection of the males depends on the acceptability of the male to the females rather than on favorable competition with other males. Again, group continuity is through the females.

Since the females provide the continuity and stability, the isolated one-male group is actually a female-centered organization to which the males are added by various mechanisms. The mother-lineage undoubtedly plays a prominent role in most of these organizations.

TROOPS

Troops of primates are social units with several adult males, several adult females, and their offspring. The social organization of troops is more complex than that in the groups above, involving an additional bond between adult males forming the adult male association. Here the adult males tend to have additional roles in maintaining and protecting the society, and they must cooperate to perform these roles effectively. The emphasis here is on protecting the group from predation, a major role that is strengthened by the form and hierarchy of troop organization. As a result, troop organization is common among ground-dwelling or partially ground-dwelling primates.

The adult male association adds an element that is lacking from the previously discussed groups; that is, the mother-lineages not only focus their activities within the kin group but also extend them outward to include the adult male association. A group of thirty to sixty animals will consist of several mother-lineages whose social interactions with the adult male association form a matrix that shifts a good part of the burden of maintaining continuity and stability onto the male association. Despite occasional additions of males from the outside, troop activity is as much male-centered as it is female-centered, and savannah baboons have clearly established the males as the determining social factor in the society.

In fact, baboons are a good example of troop organization. The central core of the troop is made up of an adult male association ranging upward from two males to as many as a dozen. Clustering around the core, both physically and socially, are adult females and their young. Farther out on the periphery, particularly when the troop is moving, are the younger males. Cooperative coalitions of adult males are so important to the survival of the savannah baboons that their social organization has emphasized the adult male association at the expense of the mother-lineages. The biennial spacing of births forces play groups outside the sphere of mother-lineages (see the chapter on play); grooming focuses on the males; and females, who do most of the grooming, lavish their attention on the adult male association (see the chapter on grooming). The result is a weakened mother-lineage with strong social bonds between females and the adult male association.

The other extreme in troop organization is represented by the Minoo-B troop of Japanese macaques. This troop is unique in that the mother-

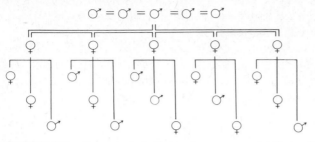

FIGURE 13   Savannah baboon troop organization with the mother-lineages restricted to two generations. The subadult sons move to a somewhat peripheral position and the daughters eventually rely heavily on their social bonds with the adult male association.

lineages are so strong that no adult male association has formed to act as the focal point for social relations. The troop has been subjugated to the mother-lineage organization to such an extent that there is no social position for fully adult males in the troop and they tend, after a short time, to become peripheral and leave. The result is that the troop is led by a female (Figure 14). More typically, the Takasakiyama troop and the Koshima troop of Japanese macaques have an adult male association that modifies the relationships between the mother-lineages and binds the society together laterally, cross-cutting the vertical bonds of descent.

TROOPS WITH ONE-MALE SUBGROUPS

Troop organization among the hamadryas baboons and the gelada is subordinated to strong subunits within the troop. But unlike the Minoo-B troop of Japanese macaques, these subgroups are male-dominated. Similar to the one-male groups described above, they are characterized by a firm social bond between an adult male and one or more adult females. Unlike the isolated one-male groups, these subgroupings are embedded within a larger social group containing other one-male groupings. As in the regular troop organization, the males have regular social relations with one another but do not form the strong adult male association found in the more unitary troops. The gelada troop seems more flexible than that

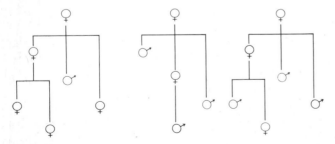

FIGURE 14   Japanese macaque troop organization with no adult male association. The males tend to be temporary members of the troop.

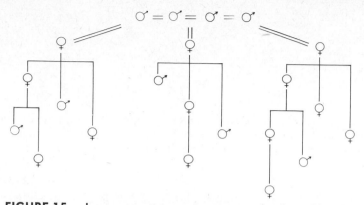

FIGURE 15   Japanese macaque troop organization with a balance between the binding forces within the mother-lineage and between them and the adult male association. Both subgroups are present and active within the society.

of the hamadryas, which has a more permanent boundary (Kummer, 1968).

The one-male subgroups in this troop organization are free to move independently of one another, and the social groups break up temporarily and regularly for feeding over sparse, near-desert terrain. Each subgroup has its own large male for protection in a region mostly devoid of large predators. This troop structure also enables large numbers of such subgroups to congregate in a single sleeping place since, even though the strong subgroups weaken the troop structure, there still are social bonds between harems that allow peaceable recongregation each evening.

The harem unit of social organization is perhaps the other extreme of a mother-lineage. For the harem system to work, the mother-lineage must be broken down completely, since the male in collecting his females removes them from their mother-lineage and breaks the bonds. Moreover, the male operating within the hamadryas troop organization does not need to displace some other male from the leadership of a mother-lineage but can build his own unit by taking over young females. Kummer's (1968) description of hamadryas harem organization shows that it precludes any effective mother-lineage organization since the male keeps his females with him continuously and through a series of threats and neck-bites prevents them from moving out into other social relationships. Here the bond between mother and infant is typically weak, and the harem leader has very little to do with his offspring.

In the harem organization, the need for social ties between one-male units is satisfied by the social interaction (about 33 percent) of the infants and juveniles with their peers outside the harem unit; by contrast, the adults experience less than one-tenth of their social interactions with outside members. In fact, the hamadryas troop is one of the most complex social units found among primates. Involving highly stable subunits whose adult members interact primarily within the harem, it is embedded

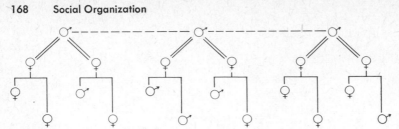

FIGURE 16   One-male subgroups with females and their offspring bound into a troop organization by the social relations between adult males. The focus of the social relations within each subgroup is the leader male.

within a troop organization that allows the younger individuals to interact fairly freely with their peers within the larger troop. It is this interaction between the younger members and between harem leader males that gives stability to the whole troop and weakens the inherently divisive tendencies of the harem units.

TROOPS WITH PERMEABLE SUBGROUPS

All of the above forms of primate social organization are permanent over fairly long periods of time, the individuals that comprise them moving together day after day. Chimpanzees have another variation on troop organization, in which subgroups fluctuate daily in their membership. Recent evidence, however, indicates that the subgroups are not simply random aggregations of any chimpanzees that happen to be passing through the area. They are all members of an organized troop in the classic manner with this difference; the troop rarely moves together as a whole but instead is broken up into smaller foraging groups that meet from time to time, join one another, exchange members, or even break up into isolated individuals only to re-form in another configuration at a later time. All the individuals of the troop (around forty to sixty individuals) know each other, and it is unlikely that a strange individual could easily join one of the small subgroups. A mother with her infant and juvenile forms one of the smallest, relatively permanent subgroupings in such a troop, and mothers with very young infants are often found together. Moreover, bands of males associate for days together and groups of mixed sexes often form, particularly females without young infants.

Chapter 8
Grooming

Grooming can best be described as a series of patterns of skin care from which certain other patterns of behavior derive. Any activity that involves going through the hair and over the skin with the hands, lips, tongue, teeth, or tail to clean out extraneous objects such as dirt, particles of sweat-salt, external parasites, or other surface materials is considered grooming. In addition, the care and attention lavished on wounds, the cleaning and removal of loose scab tissues, fall into the category of grooming behavior.

If primate grooming were confined to these processes and done only as often as necessary to achieve these ends, there would be little to say about grooming as a social adaptation and we could move on to other, more important activities. But grooming has been elaborated by primates to serve a variety of functions other than keeping the hair and skin clean and aiding in wound healing. In fact, it is difficult to demonstrate that strict skin care preceded the more elaborate patterns of grooming behavior; the functions are very different.

Primates, particularly the macaques and baboons that are so often seen in zoos and are the subject of laboratory experiments, indulge in grooming often and under many circumstances. In fact, primate species that groom frequently do so on almost any occasion that is not connected with feeding, avoiding predators, playing, or sex. In some species, sex affects the rate and frequency of grooming to such an extent that marked changes in grooming behavior follow the female's estrous cycle. In effect, many primates groom during most of their leisure time, and in zoos and laboratories they have an abundance of leisure. As a result, grooming is one of the most obvious primate behavior patterns and has been used, with various results, as a clue to many other kinds of activity. For example, over the years many attempts have been made to relate grooming with dominance, and the evidence has pointed in both directions, sometimes for the same investigator (Maslow, 1936, 1940).

Partly because grooming is not a simple, unified pattern of behavior, it must be studied within its own context, and the variations of the grooming patterns within even a single species must be considered. The most basic, self-grooming, involves no social relations and is done often, but not always, only to care for the individual's own skin. Beyond that, social grooming falls into numerous categories, each serving different ends. Relaxed social grooming, which occurs during resting periods, is an intensive activity, each partner spending many minutes at a time going carefully through the other's hair and concentrating on the grooming activity to the exclusion of other interests. In another context, grooming is a displacement activity that releases the grooming individual's nervous tension built up by the social situation when such release is not feasible through more direct activity.

THE MECHANICS OF GROOMING

The mechanics of primate grooming are determined to a great extent by the different forms of primate hands. Locomotor patterns based on climbing by grasping have emphasized the manipulative capacities of the primate extremities; in many cases, primates confront the world not only with stereoscopic color vision but also with manipulative hands. The combination of manipulation and supervision (in the literal sense) is characteristic of primate grooming activity since the latter involves close visual contact with tactile stimulation. Sight and smell are strongly coordinated, and occasionally the sense of taste is involved when an object found on the skin is conveyed to the mouth and either consumed or rejected.

The wide differences in primate hands show up in the grooming actions of various species. Locomotor adaptation is not the only selective pressure operative on the primate hand. Differences in diet have also had their effect. For example, the galago hand has adapted to a highly stereotyped whole arm control designed for striking and catching insects. As a result, when trying to catch an insect, the caged galago has great difficulty altering its strike and holding the fingers clenched to allow the hand to pass

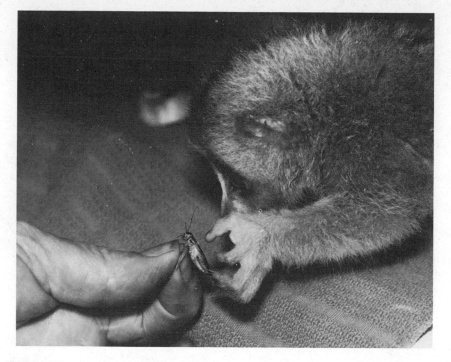

PLATE 17   A potto (*Perodicticus potto*) reaching for an insect. (*Courtesy of the Oregon Regional Primate Research Center; taken by Harry Wohlsein.*)

through the wires of the cage. Moreover, the galago hand has a precision grip for holding the trapped insect but lacks fine control of individual fingers. Since all the digits are extended or contracted at the same time, independent movement of the digits is impossible (Bishop, 1962). This whole-hand control makes it difficult for the galago to pick small objects out of the hair while grooming since the thumb is not opposable, and it uses the tooth comb instead to scrape the skin.

Since members of the genus *Lemur* are strongly frugivorous, the precision grip is not so important for them as it is for the galago with his hand full of struggling insect. Lemurs, which bring the fruit-bearing branch within reach, tend to use the hand as a hook to draw the small branches near their body. Where the surface is not hindered by hair, galagos can readily pick up objects as small as raisins by flexing their digits around the object, but lemurs tend instead to pull a hand offering raisins to them and take the raisins directly with their lips (Bishop, 1962). Lemurs also use the dental comb during grooming.

Many of the New World monkeys and most of the Old World have some degree of fine control of their digits. Full fine control means that the digits can be extended and flexed independently of one another. Minimal fine control is restricted to the separate motion of the thumb while the remaining fingers move together, separating (abducting) as they extend and converging (adducting) as they flex. Man has the extreme in fine control, in which all the digits can be moved independently of the others. Such activities as piano playing and typewriting would be impossible without full fine control. Bishop (1962) and Andrew (1963) suggest that grooming itself "could have favored independent control of the thumb in Cercopithecoidea" (Bishop, 1962), since they use their hands instead of their teeth to remove objects from the hair.

Most of the grooming activities are manual; but the lips, teeth, and tongue are brought into play from time to time. The most common form of prosimian grooming combines the hands spreading the hair apart with the use of the teeth to comb and scrape the hair and skin. The procumbent lower incisors and canines have apparently developed their special arrangement primarily as a grooming adaptation. Although this statement has been debated in the past, recent work in the laboratory and field indicates that galagos, lemurs, lorises, indri, and tree shrews all use their procumbent front lower teeth while grooming (Buettner-Janusch and Andrew, 1962; Sorenson and Conoway, 1966). Since these teeth are not used for biting during fights or for plucking fruit and only occasionally are used to scoop the pulp out of fruit, their primary function is grooming.

Among many primates, the lips and tongue are both used to clean the hair and skin. Some, in fact, suck the fur to loosen any objects tangled or stuck in the hair. Sucking and lip-smacking with tongue movements, which are often associated with grooming, could be derived from the infant's suckling movements; the lip-smacking has been further diversified as a conciliation gesture probably because of its association with the pleasurable activity of grooming (Andrew, 1963).

Even the tail is used in grooming. The prehensile-tailed howler monkeys can manipulate better with the tips of their tails than they can with their hands and as a result use the tail in manipulating the fur (Carpenter, 1934).

TIME OF GROOMING

Grooming is usually not an urgent activity. Primates groom when they are free of such pressing activities as avoiding predators, feeding, moving through dangerous parts of their range, and the like. They usually engage in grooming just after awakening in the morning, before moving out to eat, after eating, and during rest periods. After the afternoon activity and while they are moving into sleeping positions, they occupy the time with grooming. Various species time their grooming according to their different requirements.

Among hamadryas baboons, grooming is done only around the sleeping cliffs. Their movement over the near-desert regions is fairly rapid since they must cover a large area to acquire enough to eat. There is no time to sit and groom each other during the day (Kummer, 1968). The gelada have a similar problem since they, too, must spend most of the day pulling grass and feeding on the roots, a long and tedious task. Thus gelada sit about and do a lot of grooming in the morning before moving away from the sleeping cliffs; the rest of the day they spend slowly moving and eating (Crook and Aldrich-Blake, 1968). During much of the year, the savannah baboons, which have richer sources of food, groom not only in the morning but also during the midday resting period and in the evening before returning to their sleeping places (Hall and DeVore, 1965).

Factors other than the necessity for speed of movement affect grooming. Patas monkeys probably groom less when they have to move farther afield to forage, and there is a strong indication that they do not groom unless they can see over the tall grass. Their grooming is usually done when they are up in the trees or sitting on top of termite mounds (Hall, 1965, 1968). Even during slow group movement on the ground, there is little likelihood of grooming since they are not particularly secure. On the other hand, bonnet macaques, which are close to the trees and have a relatively unrestricted view, groom while the troop is moving slowly along the open roadsides. If their view is obstructed, as when a herd of cattle passes, they move into the trees.

SOLICITATION FOR GROOMING

In an earlier chapter, social distance was mentioned in connection with the spacing of individuals within the group. Few activities are tolerated that breach the social distance between animals. Typically animals register an avoidance reaction to maintain their social distance. To neutralize the avoidance reaction, an animal must somehow signal his approach when his proposed activity will involve contact with another. The literature of animal ethology is full of descriptions of the rituals that precede copulation in many species of animals. Instead of elaborate rituals, primates use a variety of gestures to perform the same function. Although grooming is not so involved as copulation nor does it put the actors in quite so vulnerable a position, it must be preceded by signals that indicate the peaceful intentions of the approaching animal.

These signals are exclusively gestural; no vocalization that could serve this purpose has been reported for any species. The facial expression during the approach is either neutral or involves lip-smacking. The animal approaches in a relaxed manner unless it is tension grooming. The more dominant the animal soliciting the grooming, the more likely it is to approach without overt gestures, using instead only neutral gestures of face and body. The more subordinate animal must often give deferential gestures if he wishes to come within the social distance of the prospective groomer. Lip-smacking and presenting are commonly used for this purpose by baboons and macaques. But usually one monkey approaches another which has often groomed it in the past and simply exposes prominently the part of the body to be groomed. If such a solicitation is not successful, the soliciting animal turns and begins to groom the other; after a time, the second animal usually reciprocates.

In some species, the presenting for grooming posture is more elaborate. Among hamadryas, "the grooming response can be released when an animal places himself on all fours in front of his partner, lowers his head and raises his tail in a gentle arch, thus presenting his side. Presenting of the rear can also release grooming" (Kummer, 1968). More often any posture that places part of the soliciting animal's anatomy before the prospective groomer and enables the solicitor to look away will suffice as a grooming signal. If the solicitor is a high-ranking animal, the response is more likely to be immediate.

SELF-GROOMING

Hall's (1962a) description of self-grooming among baboons applies fairly well to most primates.

> [Although] much of the grooming is social, . . . individuals sometimes groom themselves, while sitting, stretching out one rear limb and picking over it, or going over their chest or stomach-fur with both hands. Dog-like scratching, in standing position, is carried out by any of the four limbs. Scratching in a sitting position also occurs, notably with one hand onto the back or side of the body.
>
> When the fur has been wetted by rain, the baboons occasionally shake it off, dog-like, either in the standing or in the sitting position. They also lick the rain off their limbs. Stretching of limbs occurs, usually from the standing position, and is very similar to that of dog and cat.

Self-grooming is primarily for skin care, mostly to relieve an itch or to remove particles from the skin. But on occasions of tension, the scratching probably serves as a displacement activity. During such self-grooming, the eyes are used, not to guide the action of the hands, but to keep the source of tension in view.

Self-grooming increases with the age of the individual. Infants rarely, if ever, groom themselves and juveniles are low in active grooming frequency. Goodall (1965) never observed infants grooming themselves but saw adult chimpanzees do so frequently and for as long as fifteen minutes a bout.

A. Jolly (1966) describes the somewhat different form of lemur self-grooming:

> At intervals in the sunning trees, animals complete their morning toilet. They lick and tooth-scrape their back fur with the repeated, unmistakable forward movement of head and muzzle. They bend down and scrape their belly fur and lick their palms and their thighs. They groom their genitalia and finish with a scratch with the specialized toilet claw on the second toe or use the toe to clean their ears.

SOCIAL GROOMING

Social grooming tends to last longer per bout than self-grooming, probably because it is reinforced by stimuli from the other animal. The two individuals interact and reinforce their continued interaction by providing a variety of stimuli. This is especially true in group grooming in which a number of animals (for example, bonnet macaques) participate, turning from one to another over a period of thirty minutes to an hour.

Moreover, as Etkin reports (1963a), social grooming often serves a far broader function than skin care.

Such grooming resembles other types of persistent exchange of attentions that characterize many learned relations of familiarity in animals. The billing and cooing of doves and other birds, the persistent sniffing of one another by dogs and rodents, and the repeated prodding of the females by male guppies all illustrate the tendency common in social animals to maintain persistent contact by reiterated exchange of mild stimuli. Many social animals keep in touch with one another by repeated sounds or visual signals. Such contact presumably helps to maintain the sense of familiarity so important to their social stability.

Sparks (1967) distinguishes between three kinds of social grooming: allogrooming, mutual grooming, and reciprocal grooming. In allogrooming, one animal grooms a passive recipient. Mutual grooming is active grooming by both participants, and reciprocal grooming is the alternation of passive and active roles by both participants. Where grooming is common, all forms tend to be used, but when the members of the species groom each other rarely, allogrooming predominates. A sample of bonnet macaque grooming, taken from the author's field diary, serves to show that all three forms occur sequentially.

10 November 1961, 1203: Trixie [α♀] approaches Erra [γ♀] and begins grooming her [allogrooming]. 1204: Trixie stops grooming as Erra turns to groom Trixie [it becomes reciprocal]. 1205: Both move and Erra approaches and grooms young female A [allogrooming]. 1205½: Female A grooms Erra as Erra stops grooming [reciprocal grooming]. 1206: Trixie joins female A grooming Erra for a moment, then female A stops grooming. Then Trixie stops grooming and Erra turns to groom Trixie. 1208: Trixie moves away from Erra, then she approaches and begins to groom Snip [ε♀]. 1209: Trixie stops grooming and sits with Snip. 1210: Erra and female B walk up and sit down to groom Trixie. 1211: Trixie gets up and moves again. Erra follows. Then female C approaches and she and Trixie groom each other while Erra grooms Snip [mutual grooming, allogrooming].

Bonnet macaques may well have the most complex grooming relationships of all species of primates. The situations quoted above involved all three of Sparks' categories of social grooming and an additional one: the two females allogrooming a third female (time 1206 hours) could easily be called joint grooming or joint allogrooming. The general pattern is relatively common among baboons and macaques. Even langurs form

PLATE 18 *Papio cynocephalus* female grooming a male. Note the marked sexual dimorphism in size in these adult animals. The baboon female does the majority of the grooming.

grooming chains of three or four animals and do the kind of joint grooming described here for bonnet macaques (Bernstein, 1968).

Allogrooming is the commonest form of grooming between a primate mother and her offspring. Among many species, most of the grooming is done by the adult females, only a small amount by males and young.

FUNCTIONS OF SOCIAL GROOMING

Rat and bird husbandry have shown that touch is essential for the development of a normal personality that allows social contact between the members of a group. Members of species that are not associated by touch contact with adults during their growing period do not develop satisfactory social habits as adults. Rats that have not been handled when young are fearful and timid. If their fears are conquerable (as they are when part of the brain is surgically removed), they are still exceedingly excitable and vicious. On the other hand, those that are handled or given any other kind of strong touch sensation while growing (provided it does not involve actual injury) develop normal social behavior. As many owners of parakeets and other birds are well aware, birds that have been handled by their owners from the time of hatching are much tamer and more amenable to human associations than birds that are not handled.

This phenomenon has important and interesting implications for primate behavior. Many primates groom their infants often during the early weeks of life and continue to maintain a close relationship in later life, though usually at a lower level of activity. The contact made by grooming may stimulate the social activity of the growing animal and help it to fit into society much more easily. As has already been mentioned earlier, Harlow and his associates have demonstrated that contact is more important than nursing in maintaining the mother–infant bond. Animals without this kind of contact do not grow into socially adept adults.

COMMUNICATION

For many species of primates, grooming is the major form of tactile communication, which transmits a variety of meanings to the other members. But its primary message communicates a mood of relaxation. Marler (1965) says, "The commonest response to the initiation of grooming is a general relaxation of tension and the adoption by the recipient of postures that invite further grooming." The interaction between the two grooming animals is a series of tactile and visual signals. The initiator presents a part of his body visually to the prospective groomer; once the grooming has begun, he often shifts his position to communicate to the groomer that another area needs grooming. Likewise, the groomer pushes the groomee, who responds by shifting position so that the groomer can reach and groom relatively inaccessible areas. But the main communication is a friendly intent or at least a lack of aggressive intent.

When grooming is done under tense circumstances, the motions are quick and jerky rather than slow and careful, and the resulting communication indicates the tension of the groomer. But since this sort of grooming tends to release tensions between two animals, it can evolve into a relaxed grooming session. Most of the grooming in a species like the bonnet macaque is relaxed and rarely communicates tension.

GROOMING UNREACHABLE AREAS

Among species that do not use grooming for more elaborate social reasons, social grooming is done simply to reach areas that the individual cannot groom itself. Gorillas are the best examples of this, since their social grooming appears to be minimal.

Adult gorillas, including black-backed males, rarely groom other adults, and the grooming that does occur is concentrated on those parts of the body which the animal cannot itself reach with ease. This, together with the fact that females do not attempt to groom the dominant male or rarely groom each other, suggests that mutual grooming in gorillas has little or no social function (Schaller, 1963).

Even for a species that engages in social grooming to bind social relationships, this practical consideration may be present. Furuya (1957), for example, notes that the backs of Japanese monkeys are groomed by other members of the group far more than areas that can easily be reached by the individual. In one observation period, the back, from shoulder to iliac crest, was groomed 209 times, the forearm 8 times. However, one fact must be remembered here: since the animals often turn their backs to each other, it would be the most easily accessible area for grooming. But Furuya's figures should be kept in mind.

WOUND HEALING

The risk of danger and injury to primates is always present. Thorns and other rough vegetation often become embedded in the hands and feet and other areas. Agonistic encounters often lead to attacks in which the aggressor's teeth rip open large wounds in his victims. A narrow escape from a predator would have similar results. Any activity, therefore, that cleans wounds and removes thorns renders it less likely that a wound will fester and endanger the animal's life.

After an extremely disruptive fight in the Somanathapur troop of bonnet macaques, Slim, a subadult male, sustained a deep puncture of the muscles of his right thigh, a wide slash across the thigh, another rip wound on his left hip, and several smaller gashes. Before his injuries, Slim had associated mainly with juveniles and other subadults and was rarely groomed by adults. Three days later, however, he was observed walking along a branch to Andy, an old, relatively dominant male, and presenting his left side to Andy. The latter began to lick the left hip wound, stopping only when Slim went to work on the wound himself. Later Slim turned his right side to Andy, who again began to groom the wound. For the next few minutes, the two macaques groomed each other alternately. On several other occasions, other members that had rarely groomed Slim before licked his wounds and removed loose bits of scab. Without a doubt, Slim's rapid recovery was due to the grooming and cleaning activity of troop members.

STRENGTHENING OF THE SOCIAL BOND

When primatologists mention grooming, they usually speak of the social bond forged by this activity. Since individual animals are usually repelled by other animals, whether of the same species or not, some sort of counteracting force has to bring them together into a society. The safest way for an individual animal to escape predation and attacks by other animals is to avoid them; hence the repulsive or centrifugal tendency in most animals. However, in a society where numerous individuals cooperate for the common good and where there is a division of labor, often between the sexes, the individuals derive mutual benefit from such cooperation. The tendency to form a social bond and to indulge in an activity that strengthens the social bond becomes a selective factor in evolution. Those who do not form such bonds are more likely than others to be caught by predators.

Many activities facilitate and strengthen the social relationships between members of a society. Sexual attraction has formed societies that are usually small and incapable of elaboration on the basis of the sexual bond alone. Play also serves to establish and knit social relationships, but play alone cannot extend beyond a single generation unless the adults join actively in the play, and they rarely do. The mother–infant bond is a primary factor in the formation of mammalian societies, but in the absence of other forces that bind the lineages together, it, too, leads to a segmentary system.

Grooming is a social activity that involves members of all ages and both sexes, either as active or passive participants or both. In many primate societies, it has evolved far beyond the simple necessity of removing ectoparasites and keeping the pelage clean. By weakening the avoidance reaction, grooming emphasizes and maintains the various bonds established in the society. It gives both participants an enjoyable activity that draws them together. Alone, it cannot bind together the individuals in a group; but by bringing them into tactile communication, with the generally pleasurable message of relaxation and enjoyment, it helps to minimize their tendency to scatter.

MATERNAL GROOMING

By far the most common form of primate grooming is found in the mother–infant dyad. The primate mother begins by cleaning the newborn infant and licking it dry. The first few weeks of contact with her infant involve grooming at every opportunity, and the grooming reaches a peak. As the infant gains more independence and the mother begins to lose her almost overwhelming interest in her offspring, grooming lessens in frequency and intensity. Figure 17 shows the intense grooming during the first weeks after the birth of caged rhesus macaque infants. A similar phenomenon is found in most wild primate groups. The mother's attention focuses almost exclusively on her newborn infant, which she manipulates, examines, and grooms. Jay (1963), describing the first contacts of a Hanuman langur with her infant, writes poignantly of the mother who, when the newborn is nursing quietly or sleeping, "grooms and strokes it softly without disturbing or waking it." And DeVore (1963) asserts that

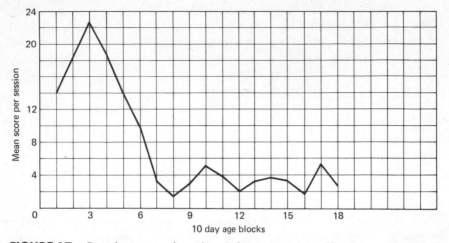

FIGURE 17 Development of mother–infant grooming. The interactions took place in an experimental cage, with the mothers having contact only with the infants but with the infants free to move to four adult female rhesus monkeys according to a predetermined schedule. *(From Harlow, Harlow, and Hansen, 1963.)*

"Grooming is the most important, time-consuming activity in baboon social life, and during the first few days of life, grooming by the mother is especially intensive. Every few minutes she explores the newborn infant's body, parts its fur with her fingers, licks and nuzzles it."

The grooming in the mother–infant dyad is heavily unidirectional. Only in the later stages of the infant's growth and development does it reciprocate, often only to solicit special favors; for example, a langur infant will groom its mother simply to get near a desirable food source that she is occupying (Jay, 1963).

For all species reported except chimpanzees, the intense grooming immediately after the birth of an infant decreases gradually until it approaches the average duration and frequency by the time the infant is weaned. Nevertheless, the relationship between mother and infant is more intense than most other relationships in a primate group, and in later life the incidence of grooming between the two is often higher than would be expected. Among chimpanzees, however, the reverse is true. During the first three months of the infant's life, the mother shows little interest in grooming it. What grooming she does tends to be rapid tension grooming rather than the slower, more effective grooming. Only as the infant gets older is the relationship characterized by grooming. The infant begins to groom its mother effectively after it is two years old and from then on their contact is in the form of grooming (Van Lawick-Goodall, 1968).

Even in a primate species like *Propithecus verreauxi,* which apparently grooms only rarely under most circumstances (Jolly—1966—observed grooming only forty-nine times during eight months of observation), considerable grooming takes place between mother and offspring. Only two species have been reported to have an insignificant amount of mother–infant grooming. Uganda mangabey mothers groom their infants less than expected (Chalmers, 1968b), and Nilagiri langurs do not engage in mother–infant grooming (Poirier, 1968).

Among other things, grooming serves to calm and relax the infant. Hanuman langurs permit other females to hold the infant shortly after birth. When the infant struggles and cries, the female strokes and grooms it; the intensity of the grooming is directly related to the intensity of the infant's distress (Jay, 1963). The calming effect of grooming is further shown during weaning. According to Jay, the only concession the mother makes during this traumatic experience is grooming. Grooming the infant before it becomes tense and aggressive calms it and postpones a weaning tantrum. Chimpanzee infants often interpose themselves between their mothers and any other chimpanzee they happen to be grooming and the mothers often respond by grooming the young intruders for a few minutes (Van Lawick-Goodall, 1967).

Any new mother–infant dyad becomes the focus of a great deal of attention during the first weeks after birth. Females, juveniles, and others that wish to look at the infant and to hold and examine it usually groom the mother first. This probably serves not only to reduce the mother's tensions and anxieties over the close approach of the other individuals but also to reward her for the expected favor of handling the newborn infant. Even the dominant females must usually groom a subordinate mother if they wish to examine or come close to the infant. In the Somanathapur troop of bonnet macaques, a very low-ranking female had the first offspring of the year. During the four months preceding the birth of this infant, she had been groomed by high-ranking females only twice. However, within the first few days after the birth of her infant, most of the females of the troop approached and groomed her. Although they were obviously interested in the infant rather than in the mother, they groomed the latter a few moments before attempting to touch the infant. DeVore (1963) reports that a highly dominant female will occasionally threaten the mother gently in order to touch her infant. But clearly the modal approach is grooming. The highly dominant male baboon seems to be the only animal in the troop that can approach without the grooming solicitation. Instead he lip-smacks (a gesture that may derive from the grooming situation), an indication of a lack of aggressive intent (DeVore, 1963).

TENSION GROOMING

The tension-reducing grooming associated with weaning and new mothers is done by one animal to reduce the tension of another whom it is trying to contact. But displacement grooming, which is characterized by rapid, jerky movements, is aimed at reducing the groomer's tensions, which are the result of imminent aggression or some other form of danger. Anthoney (1968), who describes this displacement grooming among caged baboons, reports seeing some animals that had been startled going rapidly through their own or another baboon's hair or fingering in the crevices or on the ground. Kummer (1957, 1967) and Hall and DeVore (1965) reported the same displacement grooming among captive hamadryas and wild baboons. Kummer suggests that the long hair of the male hamadryas evolved to reinforce this displacement grooming as an outlet for pent-up tensions. He also describes a female that had just received a neck-bite from the unit leader and forthwith began to groom him with short, rapid movements but without the usual removal and eating of particles that characterizes social grooming (1968).

Rhesus monkeys also show an increase in what Rowell and Hinde (1963) call "brief skin-care patterns" when the tension arising from imminent attack seizes both animals, the potential attacker and the threatened animal. Most of these reports involve caged animals; in the field, the avenues of escape from tension-producing situations somewhat reduce the importance of displacement grooming. But it is still an important pattern for some species in the wild.

Fentress (1965) found that voles used self-grooming as an intermediate pattern between freezing, fleeing, and maintenance activities. Apparently when the freezing tendency is strong, grooming does not appear until tension is somewhat reduced. But its appearance before other maintenance activities suggests again that it serves to reduce tensions to an acceptable level.

GROOMING BETWEEN SEXUAL PARTNERS

In most primate groups, the sexual bond is temporary. If this bond were the only bond reinforced by grooming, then the grooming between males and females should show cyclical fluctuations that relate to the estrous cycle. But since the other social relations within the group are rarely terminated during the mating season, sexual grooming must be sorted out from grooming related to other social interaction. The sexual bond apparently takes precedence when the female is sexually receptive, and the nature of the male–female grooming changes accordingly.

At one extreme, Sorenson and Conoway (1966) report that in several species of *Tupaia* the female is groomed by the male only when she is sexually receptive. Otherwise, most of the grooming is self-directed rather than social. On the other hand, Ploog, Blitz, and Ploog (1963) find no correlation between grooming and sexual activity among squirrel monkeys, but here again there is very little social grooming. Tree shrews use the relaxing and calming effect of grooming to establish close contact during the prelude to copulation. Among squirrel monkeys, the minimal amount of grooming that does go on is not elaborated to form a social

bond between the mating pair. In neither species is there enough grooming to consider it a bond in any other social relationship.

Many of the other primate species use grooming to reinforce numerous
social bonds, and hence one must deal with grooming and sex in terms
of changing frequencies and directions of grooming rather than in terms
of its presence or absence. Several laboratory studies indicate that the
male rhesus macaque is stimulated by the female's changing hormone
balance and concomitant increasing sexual receptivity to groom her more
frequently and more intensively as her estrous period approaches its
peak. At the same time, the female's grooming activity decreases as her
partner's increases.

Among ovariectomized females, all of the cyclical changes in the
female's hormonal balance and also the cyclical increases in male grooming were eliminated. The administration of estrogen caused an increase
in female receptivity again and in male grooming of the females
(Michael, Herbert, and Welegalla, 1966). Progesterone was also admin-
istered to the females, but the results were inconclusive; some pairs
showed increased male grooming, others did not. Moreover, even with
intact females, some males did not respond to the females' changes in
estrus. This was probably due either to an individual aberration on the
part of the male or to something about a particular female that tended to
cancel out the stimulus of the estrous changes. My observations on bonnet
macaques over many months include a female that presented for copulation many times but was never sexually mounted nor did she copulate.
For some reason I could not ascertain, the males, which readily copulated
with other females, were not stimulated to mate with her.

Many female primates have obvious signs of estrus that are easily
detected by the males; those that do not often have behavioral signs
instead. When estrus changes the female's physiology and behavior, the
male responds by coming close and attempting to copulate. Among the
species that use grooming to reduce tension, the male grooms the female
to allay the tensions generated by the close approach of a dominant male
and to form the social bond necessary for copulation. Among those species
—for example, baboons and macaques—in which the females usually
groom the males, grooming by the females decreases as the males lavish
grooming attention on them. But primates do not rely on simple stimulus-
response behavior for their adaptation. For a male that has had disagreeable experiences with a female (sexual or nonsexual) or has learned a
pattern of behavior (for example, subordination) in relation to her that
precludes copulation, the stimulus of her receptiveness will not evoke the
grooming response that usually accompanies the approach for copulation.

When the presence or absence of grooming is being considered among
bonnet macaques, not only the receptiveness of the female but also the
nature of the sexual relationship between the male and female is at
issue. The two major patterns of mating are described in Chapter 10.

Briefly, in one the male approaches and solicits copulation from the female; in the second, the female engages in a long solicitation of the male. In the first, if the male pauses to groom more than a second or two before mounting, copulation is not likely to take place. On the other hand, grooming is a significant part of the solicitation process when the female is attempting to seduce the male.

Carpenter (1942, 1964) states that both grooming and copulation increase the degree of "rapport" between the male and female, strengthen the social bonds between them, and thus prevent attacks by the male. In his words, grooming then becomes an "instrumental act" used by the female to avoid attacks by the male or to cement her relationship with him. Thus the grooming between the consorting pair reduces the female's tendency to flee and the male's tendency toward aggression.

GROOMING AS A REWARD

Grooming may be used as a reward after an agonistic encounter in which one monkey joins another as a partner. In a large troop of bonnet macaques threats are often made by more than one animal. After such a joint threat when the participants settle down to quieter activities, grooming often occurs between the partners that cooperated. The grooming is more often done by the subordinate, and although it can be partly interpreted as instrumental in reducing tensions after the agonistic interaction, it can also be regarded as a reward for support, since being groomed is a desired activity among bonnet macaques.

GROOMING WITHIN SUBGROUPS

The evidence so far indicates that grooming generally occurs between individuals that are often closely associated. Does close association stimulate grooming, or does grooming indicate and support close associations?

Within a hamadryas one-male unit, grooming activity (except for mother–infant grooming) centers on the leader male, the females rarely grooming each other. This is consistent with the social bond holding the one-male unit together. The male forms a bond with each of his females, which is maintained by the interaction between the two rather than between the females. The females compete in grooming the male, or they groom a peripheral male, but they do not often turn to each other even though they must remain close to each other as they follow the unit leader.

Among hamadryas at least, grooming appears to be predominantly a causal factor in social relations rather than a result of opportunity or proximity. Kummer (1967) reports an old male zoo hamadryas that did not groom his females as long as he had seven of them. But as soon as he had lost all but two of them, he was seen frequently grooming them. Grooming was an instrumental act used by the animal to strengthen and maintain his ties with other individuals. Grooming relationships usually indicate the direction of the bonds in the group, and those bonds are selective and hence do not radiate equally in all directions.

The one-male unit among hamadryas is one of the obvious subgroupings in primate societies. Others are the play groups, the adult male associations in the classical troop organizations, and the mother-lineages. Most infant and juvenile primates do little grooming but turn to grooming primarily in adulthood. Apparently the play behavior serves much the same function as grooming. But in many primate species, the other two subgroups mentioned above engage in a significant amount of grooming.

Yamada's (1963) analysis of co-feeding and grooming among the mother-lineages of the Minoo-B troop is another example of the binding forces exerted by grooming within and between subgroups of a society. Yamada found a higher frequency of grooming among individuals related through kinship than among others. The two most common grooming relationships are mother–child and sibling, in that order. Another common form of grooming is between individuals that are not related but have a strong "repulsive" relationship. Yamada's grooming data differ from his information on co-feeding since only those that are related through kinship can feed close together in a relaxed manner, while non-kin occasionally groom one another.

Sade's (1965) analysis of mother-lineage grooming among the rhesus macaques of Cayo Santiago indicates that a high percentage of grooming there occurs between kin (mother–offspring and siblings). He found that the founding female's infant turned to its older siblings and their offspring to establish strong grooming relationships after the mother's death. Thus, after her death the focus of the mother-lineage shifted and the grooming relationships adjusted to maintain a new set of relationships.

In most macaque and baboon troops and among various arboreal and semiarboreal forms like the howler monkeys and the vervets, the adult male association is central to the troop organization. Here grooming does not seem to play an important part in binding together the adult male association since grooming between males is often strictly limited or nonexistent. Among bonnet macaques, however, grooming is an all-pervasive social activity. During 700 hours of observation grooming occurred among the ten adult males 186 times, 51 of which occurred among the four top-ranking males. Such close and regular contact between these males is bound to influence their social relationships.

GROOMING AND SOCIETY

Virtually everyone who has discussed grooming has mentioned its binding force in society but only rarely has anyone described how grooming functions as a cohesive social force. It is commonly assumed to reduce the tensions between the two animals concerned and to supply a pleasurable activity to both participants.

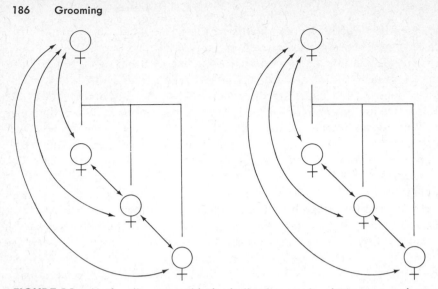

FIGURE 18   Mother-lineages with the bulk of grooming between members of the lineages. The Minoo-B troop adds grooming between lineages but does not focus this grooming on a leader or an adult male association.

We can distinguish three kinds of grooming: (1) relaxed social grooming between individuals that already have a strong attractive bond between them; (2) soliciting grooming, which allows the individuals to remain close for a temporary activity; (3) tension grooming, which serves to deflect an imminent attack. The latter two forms undoubtedly forge a temporary bond between the two individuals involved, whereas the former strengthens an already established relationship.

Relaxed social grooming begins with the mother–infant relationship. As the infant grows to adulthood, the bond between them continues to be strong, and female offspring especially tend to return their mother's grooming attentions and thus change an allogrooming relationship into a reciprocal and possibly a mutual one. Anthoney (1968) describes how this process is initiated among baboons. He asserts that some infants, if not all, learn that when they are being rejected by the mother during weaning they will be permitted to remain near her if they groom her. Thus, the frightened or insecure infant compensates for the loss of its mother's breast by grooming her. Under similar circumstances in later life, it is natural for it to groom other baboons as well.

Since an older female often has young of several ages with her (an infant, a juvenile or two, perhaps two or three adult daughters with their offspring), her matrix of grooming relationships tends to be multi-directional. Because of the bonds that hold all of these individuals close to the old mother, they are usually in close proximity to one another. Grooming between them is common, especially among macaques, which seem to maintain fairly strong mother-centered lineages.

Among baboons, for whom troop organization is more crucial for sur-

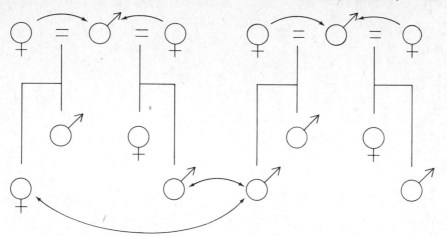

FIGURE 19 Grooming in hamadryas one-male units. The male's females focus their grooming on the leader rather than on their offspring, leaving the latter to form such grooming bonds in the formation of a new unit.

vival, the situation differs somewhat. The mother-lineage is subordinated to the adult male association as the focus of social relations (Figure 19). Grooming is still common between females of savannah baboon troops, and studies conducted for many years might show that much of it is between kin. However, much of baboon grooming is between the female and the male, the female doing most of it. Such grooming occurs characteristically when the animals are relaxed, after feeding, during a midday resting period, or before moving into the sleeping trees. It is the sort of grooming that occurs in a long-established and stable relationship, strengthening the bonds of the individual female to the adult male association without particular reference to her mother-lineage.

Grooming occurs between adult male hamadryas baboons before they are established as unit leaders and after they have passed the age beyond which it is difficult for them to maintain a harem (Kummer, 1968b). Under these circumstances, grooming is a means of maintaining bonds among the males and would be a significant factor in holding a hamadryas herd together if it were not for the fact that they cease these intermale grooming contacts when they become unit leaders. As a result, the vertical divisions of hamadryas society based on harem groupings supersede and subordinate all other social units for the adult and fully participating harem hamadryas baboons. Female grooming, like that of the savannah baboon females, centers on the unit leader. But grooming is minimized between the adults and their offspring, a fact that tends to weaken the bonds between them and to enable the young adult males to form strong bonds with juvenile females by caring for and grooming them.

Solicitation grooming is found in the more casual relationships that, within the total group, can be considered less intensive. Since the sexual relationship between male and female is, in primates, basically a tem-

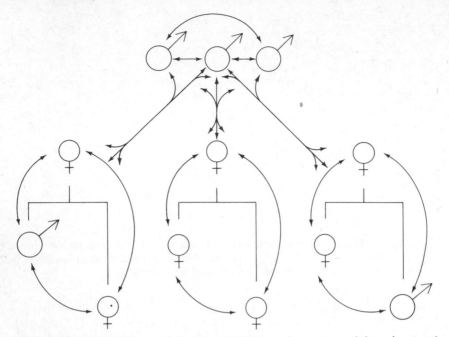

FIGURE 20   Mother-lineages focusing their grooming on an adult male association. For bonnet macaques, the whole matrix of grooming is exploited.
Baboons emphasize those relationships between the females and the males of
the adult-male association but minimize grooming between the adult males.

porary one, grooming is merely part of a sequence of events leading up to
copulation. There is, in effect, a goal, and grooming is used to attain that
goal. The same can be said of the grooming done by females and juveniles that are attempting to get a close look at the newborn infant and to
handle it. In contrast, relaxed social grooming is probably an end in
itself with no further goal than, perhaps, acquiring the reciprocal attentions of the groomee. But if Anthoney (1968) is right, relaxed social
grooming may begin developmentally as solicitation grooming by an
infant seeking comfort from its rejecting mother.

Tension grooming appears capable of deflecting an attack or of reducing tensions after one. Hamadryas females groom their males after the
latter have bitten their necks to force them to follow. It interposes a
pleasurable activity between the two and reduces the tensions. This
form of grooming is as capable of reinforcing social relationships as the
relaxed social grooming found in the mother-lineage. Although most of
the grooming between a hamadryas unit leader and his females is probably the relaxed form, the tension grooming serves during stressful interactions to reduce the disruptive tendencies and allows the baboons to
return to a lower level of arousal without the female's having to escape
from the vicinity of her male. Such an escape would tend to break the
bonds holding the one-male unit together.

Tension grooming may be one of the few binding forces between mother-lineages in the Minoo-B troop of Japanese macaques. Yamada (1963) mentions that there are several forms of grooming but does not distinguish between them. If the Minoo-B troop's nonkin grooming is primarily tension grooming, it would fit the pattern of reducing tensions and binding together social relations that have strong repulsive tendencies and, as in the one-male hamadryas units, reduce the divisive tendencies of the group.

In the relaxed social relationships of the gorilla group, where social boundaries are far less distinct than they are for baboons and macaques, the shifting males do not form strong bonds with either the silver-backed leader male or with the females. Male–male grooming is nonexistent and male–female grooming is virtually so.

Thus primates have elaborated a relatively simple skin-care pattern into a hierarchy of social binding forces that can both strengthen the bonds formed on the basis of attraction and reduce the divisive forces in a society of aggressive individuals. Grooming tends to be far less significant among tree-living primates and among the larger primates that have fewer predators. Thus grooming can be used by primates that require strong binding forces to overcome aggressive tendencies that threaten to divide their society. But grooming is only one of the behavior patterns that can be so used.

Chapter 9
Play, Exploration, and Socialization

Although play, exploration, and socialization can be considered as separate topics, they are, in fact, interrelated. The behavioral sequence that begins as exploration may during its course become play and part of the individual's socialization. One of the products especially of social play, socialization can be considered a long-range goal of play.

DEFINITIONS

Play is one of those activities that nearly everyone can recognize but almost no one can adequately define. Different observers usually agree about whether a particular behavior sequence is play or not; but since the same general patterns of behavior that characterize play are found in fighting, copulating, eating, and other immediately goal-oriented behaviors, play cannot easily be distinguished from the others by a simple description of the activities themselves. *How* rather than *what* distinguishes play from nonplay.

BEHAVIORAL SEQUENCE

The sequence of activity patterns is a major cue to play behavior. Play can thus be defined as reordered, repeated, fragmented, exaggerated, and incomplete patterns of behavior that are otherwise directed to such ends as feeding, sex, predator avoidance, locomotion, and any other immediately goal-oriented behavior (Loizos, 1967; Marler and Hamilton, 1966). Only the intention movements that initiate play or the meta-communication about play are unique to play; and in the repertoire of some species, even they can be considered the greeting behavior that leads sometimes to play, sometimes to grooming or to sex.

The basic characteristic of play is that it does not lead efficiently from an appetitive behavior (seeking or preliminary activity) to a consummatory behavior (such as orgasm or eating). The behavior patterns incorporated into a play sequence are drawn from a mixture of other activities in a seemingly random order to produce a sequence that appears goalless. During play, gorilla infants use some of the elaborate chest-beating display—beating on their chests, slapping at foliage, and holding a leaf in their lips (Schaller, 1963). Bonnet macaque infants shift rapidly from chasing to mounting to chewing on a leaf and then back to chasing. The chase does not culminate in an attack, the mounting does not lead to intromission, and the chewing does not lead to full mastication and swallowing. But each is repeated many times as the animals continue their play.

The second major characteristic of play is that the movements are exaggerated. Rocking gaits are used instead of the smaller and more efficient forms of running. Grappling motions, derived from attack patterns, are elaborated until they are inefficient. A leaf is carried in the foot and waved about rather than carried inconspicuously in the hand.

The fragmented, incomplete, reordered, and repeated behavior patterns used during play give it its nondirected character, and the exaggerated movements clearly indicate that it is not meant to be an efficient means of attaining the goals that are otherwise associated with the patterns.

LEVEL OF AROUSAL

Play is sometimes defined in terms other than the movements or patterns of behavior that are used in play. Mason (1965b) suggests that play can be defined partly in terms of the level of arousal the animal shows at a particular time. There are indications that to begin play, the individual must be aroused to a level above the minimum, that play itself augments and raises the level of arousal, and that too high a level of arousal inhibits play.

Thus Mason found that pairs of chimpanzees thrust into three different situations that differentially changed their level of arousal reacted by altering their proportions of playing and clinging. In their living cages where arousal was lowest, their clinging responses constituted 15 percent of the total clinging and playing responses. In a holding cage that was moderately familiar to the animals but was not a regular part of their daily experience, they clung 23 percent of the time and, finally, in an unfamiliar room where their level of arousal was high, they clung 51 percent of the time. This indicates that the higher levels of arousal inhibit play.

On the other hand, novelty, which produces a higher level of arousal than familiar surroundings, may enhance play if it is not too strange. Using six pairs of wild-born chimpanzees, Mason (1965b) observed that "play was consistently more frequent in [a] moderately unfamiliar situation." He also found that play increases the level of arousal, probably because the mixed and exaggerated behavior patterns during play stimulated the various senses. As the level of arousal increases, an animal makes fewer contacts with novel stimuli. This phenomenon has a corollary that has been regularly reported in wild primates: As play becomes more intense, younger animals tend to drop out of the play groups. Apparently, as the individual arousal level increases with prolonged play, the lower threshold of fear prompts the younger animals to return to their mothers or to drop out for a while to reduce their level of arousal. In fact, Harlow and his associates have shown that even for play to begin, each animal must be able to lower the arousal level below the fear threshold. His infant rhesus macaques, raised on cloth surrogate mothers and faced with a novel stimulus, first clung to their "mothers" and then moved out to examine and play with the new object. Those without recourse to clinging could not overcome their fear, that is, reduce their level of arousal to investigation and then play levels, and thus remained in a fear huddle.

EXPLORATION VERSUS PLAY

Expl. occurs at higher arousal levels.

According to Hutt (1966), among human children, exploration is the first reaction after fear to new situations and new objects in the environment. Children who are introduced to a novel situation by themselves and who are thus without a quick means of reducing their arousal level have neophobia. After the object or the situation has been fully investigated by the human child and is familiar to him, play can begin. But the play and the initial exploration differ. Exploration is goal-oriented and its aim is to find out about the new stimulus and become familiar with it. To accomplish this, the human child focuses all its senses, particularly those of vision and touch, on the new stimulus and carefully examines all aspects of it. Only after the child has learned all the inherent possibilities can play begin. During play, the senses are not concentrated as they were during exploration, and the playing child spends some of its time manipulating the object while looking elsewhere. Play is thus more relaxed than exploration, which tends to be an intense focusing of interest.

Minimal arousal is for resting, low to moderate arousal is necessary for play, moderate to high arousal results in investigation, and high arousal produces fear. Play occurs in known circumstances, usually in regular patterns and in regular places that are secure and safe. In the laboratory, a bare cage does not, after it has become a familiar setting, produce sufficient arousal for play. In the wild, the stimulation of other members of the play group in a familiar setting is probably sufficient to raise the level of arousal to the point necessary for play. Thus bonnet macaques play in the same setting of intertwining branches day after day and apparently do not require a new setting to arouse them to play. Novel stimuli, such as a new windup toy in the laboratory or the approach of a strange animal in the wild, produce such a high level of arousal that fear is the first response; if the level of arousal can be reduced in the presence of the stimulus, the fear changes to curiosity about the new object. Play comes only after these earlier responses. An attacking predator, of course, does not allow the level of arousal to diminish to the point of investigation, much less of play.

Ontogenetically, exploration appears before play in infant primates. Harlow and Harlow (1965) postulated a stage of exploration before interactive play. "Early exploration [in rhesus infants] consists of brief sessions of gross bodily contact and oral and manual manipulation of inanimate and animate objects." There is no reliable laboratory evidence that object (or environmental) investigation and exploration significantly precede social exploration, and observations in the wild are no more conclusive since young infants seem to explore both animate and inanimate aspects of their environment indiscriminately at first. Early in life, the primate infant in the laboratory prefers social exploration to object exploration. Although this may be partly due to the responsiveness of cage mates, who keep the level of arousal high in contrast to the impoverished cage surroundings, it probably indicates the necessity for most primate species to develop fairly elaborate social relations.

Exploration is, in part, the process of becoming familiar enough with the environment to enable the young infant to move away from its mother without extreme fear. The level of arousal during the first moments of awareness of the world is too high to allow for play. All the senses are concentrated on the immediate environment, animate or inanimate. The infant's first vantage point for exploration is its mother's breast. Then, as its strength and coordination increase, the infant begins to manipulate objects with its hand, running it through the dirt or feeling the leaves and fruit on a tree. It smells many objects and tastes some as well.

Very early in life exploration begins to give way to play. Watching a young monkey discover a new object, one is struck by the similarity between it and the children described by Hutt. At first, the infant focuses all its senses on the object, picking it up, turning it in its hands, holding it up to sniff at it, looking at it closely, and perhaps nibbling it. But once the object has been smelled, felt, seen, and tasted, the infant is likely to look around at its surroundings, meanwhile idly manipulating the object in play. Part of the process is seen in Jolly's (1966) description of captive *Lemur macaco* and wild *Propithecus verreauxi.*

. . . the first of all games is to leave the mother, crawl off two steps, then dash back, landing on the mother. Then the infant crawls two or three steps and dashes back, then three or four steps and back. If the mother sits still, the infant at last may move a meter away, usually upward, so that it drops down with some force and speed to its mother's fur. The whole procedure is an obvious game, from the slow adventure into the unknown to the delighted, reassuring bounce onto the mother's fur. At any movement from the mother, the infant scrambles toward her with frantic contact calls, legs and arms windmilling ineffectively.

Although Jolly considers the whole process a game, the slow, tense movement away from the mother is probably more akin to exploration, whereas the "delighted, reassuring bounce" has the hallmarks of play.

Exploration continues as the infant grows older, but the rapidly increasing frequency of social play in most species of primates reduces exploration to a small fraction of the total behavior. The early exploration involves the infant's process of learning about its environment as a whole. Later exploration is simply additive, focusing on new elements of the environment as they appear and learning about them against a background of known elements.

FORMS OF PRIMATE PLAY

The usual division of play is into solitary and social. In terms of the level of arousal necessary for play to take place, there is probably no difference. In solitary play, the individual responds to the slightly novel aspects of its surroundings that are inanimate; in social play, it is stimulated by animate beings. Social play undoubtedly provides a greater amount of feedback, since the participants respond to each other. I know of no recorded cases in which solitary play was broken off because the level of arousal has been raised to the fear level by the activity alone, but social play often stops momentarily while the participants rest and reduce their level of arousal. Perhaps because solitary play does not maintain the necessary level of arousal, it tends to end in boredom.

SOLITARY PLAY—LOCOMOTION

Locomotor play is generally solitary, partly because of arbitrary distinctions. Certainly social chasing is a form of locomotor play, but it is primarily social and only secondarily locomotor. Even the generally accepted solitary locomotor play can become social when two or more young infants hop in the vicinity of one another. But such hopping is not usually directed toward the other participants, and the only social aspect is the reinforcement they receive from each other to continue.

Much solitary locomotor play involves hopping about on the ground or in the trees, climbing and dropping in the branches and hanging from one or more hands or feet. The primate species that eventually engage in much social play generally devote only a small part of their activity to solitary play, which does not become very elaborate. Animals isolated in cages are more likely to extend the period of solitary play than those in the wild. Thus under favorable conditions, locomotor play can become far more elaborate than that described above.

Tupaia may invent fairly complex locomotor play, such as back somersaults. They have strong tendencies to stereotyped activity. The pattern of this activity changes, however, from day to day, and tends to grow more complex and to incorporate longer and more "difficult" leaps, within the limits of the cage. . . . Lemurs indulge in a great deal of locomotor play. They can travel upside down, hang by their hands or feet, and ricochet around a familiar room using only the most minute handholds (Bishop, 1964).

Describing gorilla play, Schaller (1963) writes: "Most lone play involves climbing and swinging, jumping and sliding, waving of arms and legs, batting of vegetation, somersaults, and running back and forth with exaggerated gestures."

Because most species of *Tupaia* live in small social groups with no more than one adult of each sex, there is probably a high frequency of solitary play. Schaller reports that 43.4 percent of his observations of gorilla play were of solitary play. The lemurs reported by Bishop, as well as the *Tupaia,* were caged animals. Free-ranging bonnet macaques, baboons, langurs, and many other highly social primates engage far less in solitary play and do not elaborate it so much. Because of their propensity for social play, they are stimulated by the sight of a young monkey playing alone to join it and thus to turn to the patterns of social play.

SOLITARY PLAY—OBJECT PLAY

Among most primate species, play with objects can be extremely complex because of the opposability of the thumb, which enables them to manipulate objects fairly efficiently. Hence much of the laboratory research on primate behavior has focused on establishing primate mental abilities through the manipulation of various objects and tests. This has led to the rather erroneous impression that primates manipulate their environment a good deal. As a matter of fact, very little of their play or general daily activity involves the manipulation of objects except to prepare food and convey it to the mouth.

Occasionally objects become the focus of social play. In a two-acre enclosure at the Oregon Regional Primate Research Center, Japanese macaques use any stray object as an excuse to chase the individual that has it. Ten or twelve infants and juveniles converge on the animal and attempt to wrest the object from him. But this group is living in an impoverished environment since, despite its size, the enclosure offers very little variety. Bonnet macaques in the wild occasionally wave a leaf in one foot as an invitation to play, but the leaf is dropped as soon as the chasing and wrestling begin and the leaf does not become the focus of play.

SOCIAL PLAY

Social play in primates occurs between individuals of all ages and both sexes. But like their human counterparts, it is primarily the young that play with one another. Species differences in the amount of adult play and the amount indulged in by each sex will be discussed with the function of play.

For many growing primates (gorillas are a major exception), play fills the time between eating and sleeping. For example, howler juveniles spend 80 percent of their waking time playing (Carpenter, 1965). Most of this social play consists of wrestling and chasing. In laboratory rhesus monkeys, Harlow and Harlow (1965) describe wrestling as the simplest form of interactive play, "in which infant monkeys wrestle, roll, and sham bite each other without injury and seldom become frightened." They describe chasing as a separate pattern of noncontact or approach–withdrawal play. Both appear at the same time in rhesus macaque infants, but among bonnet macaques there is some evidence that the approach–withdrawal play chasing begins earlier than wrestling. Among them, chasing is definitely a milder form of play and is characteristic of the smaller infants. As the bonnet macaque infant grows, it participates more in larger play groups in which wrestling is the main feature. When the wrestling becomes too violent, smaller infants and females, infant or juvenile, drop out of the group. Although chasing and wrestling continue to be part of the play activity of older juvenile and subadult bonnet macaques, wrestling predominates from the second year on.

Tail-pulling is another major sport among species with tails sufficiently long to make it interesting, for example, longer-tailed macaques, langurs, and baboons. Infant bonnet macaques chase each other up and down the aerial roots of banyan trees, and often during the chase one infant grabs the tail of another and swings from it until both drop. Sometimes a chain of three animals, the lower hanging onto the tail of the one above, will swing for a moment from the aerial root.

When water is available, swimming is common among macaques. The Somanathapur bonnet macaques were observed chasing each other and playing hide-and-seek in the shallow water. Some even disappeared under the surface and swam several feet submerged, only to leap out and surprise another juvenile. They occasionally slid down slippery banks into the water. The rhesus of Cayo Santiago make spectacular leaps into shallow pools from high branches during their play in the water.

Swimming is one of the best examples of the differences in play patterns that derive from the abilities and adaptations of the various primates. None of the brachiating apes do any swimming because their shoulder girdles are strongly altered from the quadrupedal condition found in most other primates. When they fall into the water accidentally, their normal arm movements are not oriented in a direction that will keep them afloat or move them through the water. In contrast, the quadrupedal monkeys simply dog-paddle around, using their usual locomotor patterns. Human beings likewise must be taught new motor patterns to swim, since they share their range of arm motion with the apes rather than with the monkeys.

Patas monkeys, which are the only monkeys truly adapted for running on the ground, play in elaborate, long-distance chasing. "The surplus energy of the young patas is almost entirely spent in social play interactions which involve a most vigorous exercising or practising of speed and agility of ground locomotion which can be readily seen as an adaptation for survival from day-hunting predators such as cheetah or hunting dogs" (Hall, 1965). Monkeys with prehensile tails hang by their tails while wrestling in the branches, whereas Old World monkeys without such equipment hang by their feet and hand-wrestle instead.

Some of the patterns of sexual behavior occur in play. Mounting is very common among baboons, macaques, and vervets. The mounting animal occasionally gives pelvic thrusts while mounted but intromission during play does not occur. Gorilla infants sometimes go through a somewhat ritualized arm-waving during play that is similar to the precopulatory behavior observed between adult gorillas at the Columbus Zoo (Schaller, 1963).

There are numerous variations on the basic chasing and wrestling pattern, but for many primate species the patterns seem rather stereotyped. The participants chase through the branches or, if on the ground, around the bushes and other landscape features in a thoroughly predictable manner. Even the elements of play can be put together in the same manner time and time again. Chimpanzees offer a striking contrast to this pattern since even in the wild they are more inventive about using their environment to develop elaborate games.

COMMUNICATION IN PLAY

Obviously, individual play does not require communication among the members of the group, but social play, like all social activities, generally requires communication signals. The first of several communication prob-

lems to be solved by playing primates is to indicate that the social interaction is play rather than some other kind of activity, particularly since
play often includes elements of aggressive behavior. Second, adults need
to control the play group to prevent it from becoming disruptive, and
third, many species use a signal to reinforce the continuing play for the
participants.

Altmann (1962) describes the signals used by primates to indicate play
as "metacommunication," or communication that modifies the meaning of
the communication to follow. He feels that these metacommunication
signals change the responses of the receiving individuals from a potentially apprehensive to a playful reaction as long as they clearly indicate
that play is meant. Once the signal becomes ambiguous, however, the
play intent can be mistaken for the other behavior patterns from which
the elements of play are drawn.

Two kinds of metacommunication are therefore probably necessary.
One set, to initiate play, must clearly communicate that the sender is in
a playful mood and that all of his actions are to be interpreted as play
rather than aggression, sex, and so on. Another set must continually reinforce the fact that the activity is play. The same signals can be used for
both purposes, and usually the initiating signals are repeated whenever
a clear restatement of purpose is necessary. Other signals, mainly weaker
reinforcing signals, have been reported, but no experimental data are
available to prove their function.

Despite specific and generic variations, intention movements to initiate
play seem to conform to a regular pattern. The most often reported, the
"play face," is probably the clearest play signal among primates, since
much of their communication is in facial gestures. The chimpanzee play-
face, for example, differs from the fear grimace and the open-mouth
threat by exposing only the lower teeth, whereas the fear grimace pulls
the lips back tightly to expose all the teeth and the open-mouth threat
does not show any teeth. Bonnet macaques use a play-face that is also
intermediate between the fear grimace and the open-mouth threat, but
with them the upper and lower teeth are shown slightly in the play-face
but completely exposed in fear and completely covered in threat.

The kind of locomotion used during a play approach also communicates the individual's intent. In bonnet macaques, the "play gait" is a
fore-and-aft rocking gait, a relaxed, loping gallop that contrasts strongly
with the tense, efficient running during an attack. Van Lawick-Goodall
(1968) describes the "play walk" of chimpanzees: "In play walking, the
chimpanzee walks with a rounded back, head slightly bent down and
pulled back between the shoulders, and takes small, rather 'stilted' steps.
Often there is a pronounced side-to-side movement as he moves forward,
rather like the seaman's roll." She has also seen the gamboling play-gait,
similar to that of the bonnet macaques, used as a play approach, and
Hall (1967, 1968b) describes a similar play-bounce for patas monkeys.
Whereas the play-face is directional (aimed by the sender to a particular

receiver), the play-gait is a signal that can be received from any angle and by the group in general.

In addition to the above overt signals of intent to play, play can also be initiated by what can be termed a broadcast invitation to play. Carpenter (1934) described such an "invitation" given by a howler monkey juvenile that was swinging by its tail and then began to thrash the air with its legs. Soon it was joined by another young monkey and the two began to wrestle while suspended. Within a short time, six other animals had joined in the play. Whether or not this activity can be construed as an "invitation" to play we cannot, of course, be sure. But Carpenter discreetly observes: "This is a good example of play between two animals serving to stimulate similar responses in other associated individuals." I have seen bonnet macaque infants and juveniles in a similar situation picking up a twig, a leaf, or their own tails in one hind foot and gamboling along the ground, thus stimulating other infants and juveniles to join in social play.

The generalized greeting embrace characteristic of gibbons does not specifically indicate play since it leads to grooming and copulation as well (Carpenter, 1940). It does, however, indicate that what is to follow is friendly.

During the actual play, the rapid alternation between various patterns of behavior (sex, aggression, grooming, and eating) probably serves to continue the play mood. Since no single activity predominates, the individuals do not fall into the mood of that particular activity. If the alternation is not frequent enough, there is danger that play will cease and some other activity will replace it. For example, during play bonnet macaque juveniles and subadult males frequently concentrate on wrestling. If the wrestling becomes too intense and is not broken by other patterns such as chasing or mounting, play degenerates into aggression and one or more of the participants breaks away and shows fear grimaces.

Another reinforcing signal is the low grunting often associated with play. Although most play is silent to observers beyond a few yards, bonnet macaques, gorillas, and baboons emit low grunting sounds or, as Schaller (1963) describes it, "a very soft panting chuckle (a-a-a-a)" that is expelled through the mouth. Since these sounds are associated only with playful activity, they probably communicate a playful mood and thus reinforce the continuing play interaction. Ploog (1967) reports for squirrel monkeys that "Play-squeaking always accompanies playing, chasing, grasping and playful nipping as well as preparatory sex play. Since we never observed that playing ever changed to fighting, in spite of the aggressive elements characteristic of this type of behavior, we believe that some sort of signal is involved here which tells each of the participants that there is no aggressive intent (sounds of assurance)."

Adults also control the play of the younger members by signals. Howler males use vocalizations. Carpenter (1964) states: "That is particularly true of play fighting when one of the young ones gives a cry of pain.

When this occurs, almost invariably a near-by male will vocalize by a series of grunts and the quality of the juveniles' behavior changes immediately." A noisy baboon play group almost invariably brings a reaction from an adult male, who interrupts their play with a grunting threat or approaches the play group; the approach itself communicates imminent aggression.

MOTIVATION VERSUS FUNCTION IN PLAY

Many authorities, puzzling over the function of play, have concluded that play provides future rather than present advantages. But the question of motivation remains. Why do primates (or any other animals) play? Apparently the young monkey is not motivated to play to increase its motor control for running or climbing or to become properly socialized. But the individual must approach play with some immediate motivation; otherwise it would not play. Thus, whatever the long-term functions of play, the immediate cause of play remains ambiguous.

Unfortunately, studies on play motivation are rare and inconclusive. The most that can be said now is that the motivational factors must be complex. For example, the need for activity could motivate the animal when he has reached a certain level of arousal. But if so, why do female primates play less and groom more than males of the same age? The sex differences suggest some kind of hormonal influence. Again, the enjoyable social interaction and physical contact of play probably supply an immediate goal that stimulates young animals to play. But then how do adult animals satisfy such requirements when play no longer satisfies the need for them? In other words, how do changes in the hormonal or other physiological systems eliminate the need for play?

During the following discussion, we attempt to present some answers to the questions above, aware of the inherent danger of imputing anthropomorphic motives to the playing primates.

FUNCTIONS OF PLAY

All of the usual theories about the functions of play relate play to future goals rather than to immediate needs. The issue is further complicated by the fact that caged and free-ranging animals differ markedly in the amount of play they engage in. For young caged mountain gorillas, play is reportedly "the most prominent modality of activity" (Carpenter, 1937); for their counterparts in the wild, Schaller (1963) reports that play is relatively rare and nearly half of it is solitary rather than social. Eisenberg and Kuehn (1966) note that captive preparturient *Ateles geoffroyi* females play at an age when wild females have long ceased to do so. Field and laboratory reports on other species support the conclusion that one of the immediate functions of play is to occupy the extra time made available to captive animals, which are cut off from much of their normal daily activity. As a matter of fact, play, like grooming, is a leisure-time activity rather than a maintenance activity such as feeding.

FUNCTIONS OF SOLITARY PLAY

Since solitary play is here distinguished from exploration, only the manipulation of familiar objects and locomotor play qualify as solitary play. One function of manipulatory and locomotor play is practice and experience in motor behavior. Such exercise ensures that the individual has the skill to escape a predator. Otherwise, the endurance and speed in running and climbing are unnecessary in day-to-day activities.

Preparation for future activities may also be part of the function of solitary play. One good example is the nest-building play of infant chimpanzees long before they are old enough to need their own nests. Van Lawick-Goodall (1967) describes the baby chimpanzee Flint's pushing down grass stems and sitting on some in the proper nest-making manner but shoving others in his lap. The motor experience here is not necessary for later life since captive chimpanzees easily learn to build nests once they are exposed to materials and another nest-building chimpanzee. But it is one way of developing skills at a time when mistakes will not be dangerous or disastrous.

FUNCTIONS OF SOCIAL PLAY—THE SOCIAL BOND

Play as a social instrument brings the playing individuals into close, intimate contact that often involves intense tactile stimulation. Repeated play, day after day, enables the regular playmates to maintain a group familiarity and a habitual bond that persists beyond infancy and is usually reinforced in later life by other processes of familiarization, for example, grooming in primates. "Play may thus be considered one of the techniques of reiterated stimulus exchange by which social animals maintain their familiarity with each other as individuals. It appears to be an important technique by which young social mammals learn their place in the group and develop appropriate in-group feeling. It is often conspicuous in maintaining pair relations in social mammals" (Etkin, 1964b).

If play is one of the means of forming and maintaining social bonds, then it should appear in different subgroups in the various primate social structures. Bonds that are necessary to the group's adaptation develop through the use of play between selected individuals. Even with the meager evidence available today, a comparison of the nature, composition, and size of play groups shows that play is used selectively and reveals a great deal about the society in which the playing animal lives.

Differing Social Structure

We have seen in the earlier chapter on social organization that primate societies modify a few basic principles to produce different social structures. The following examples will show how play is used in various ways to form particular social bonds.

Baboons   Savannah baboons have one of the most tightly knit societies. Although the group is relatively impermeable, strong social bonds must be formed to offset the centrifugal force generated by the high level of aggression needed to protect the group from predators. These fundamental social bonds are developed in part by the play of infants and juveniles.

Because baboon births are spaced nearly two years apart, the infants do not have siblings close to their own age. By the time the infant is beginning to play significantly with others, its next oldest sibling is two and a half years old and has established strong relationships with its peers. Even if the mother lineages were strong in baboon society and cousins were available for play, only a few young would be of the same age in their kin group. Thus the young savannah baboon seeks playmates outside its kin group in an age-grading system reminiscent of many human societies in East Africa. These age-graded play groups, like the human ones, form social bonds outside the lineage system that cut horizontally across the vertical kin relationships, formal in human societies, informal but nonetheless permanent in baboon society.

Figure 21 shows that few play groups can be formed within a mother-lineage. Only for year six are there two males of the same age in one mother-lineage, and since males do much of the rough-and-tumble play, the number of males is of greater consideration after the first year than the number of available females. If A3 and A4 formed a play group, the likelihood is that C4 would be a regular part of that group. Moreover, A8, B6, and perhaps C7 would form another play group. Certainly by the time A8 was born, A3 and A4 were no longer playing gentle chasing games; and when A8 reaches the age of rough wrestling play, his elder male kin are fast becoming adult males. Population dynamics alone are enough to force the formation of play groups that cut horizontally across the lineage relationships.

The peer group becomes an increasingly important focus of the growing baboon's activity. "Throughout the first year, the infant becomes progressively more attached to this peer group and more independent of the mother. Finally, it is the infants of the previous year, now largely independent of their mothers, who play the major role in enticing the young infants from their mothers and initiating them to group play" (DeVore, 1963). Thus in Figure 21, C3 would draw A3, A4, and C4 into social play, although only C4 belongs to the same mother-lineage as C3. Since the infants of the previous year are not old enough to be siblings,

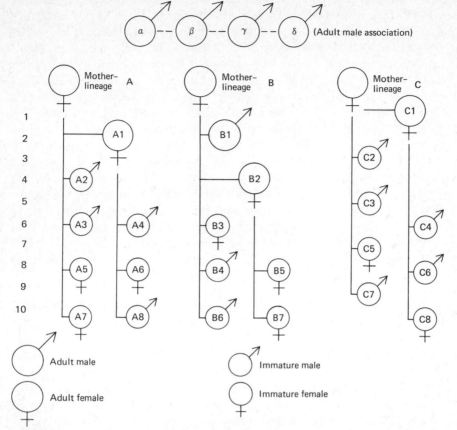

FIGURE 21 Diagram of a troop organization with mother-lineages indicated. The births are spaced in two-year intervals, as is often the case in savannah baboons.

they draw the younger infants into social relationships outside the kin association.

Baboon play is frequent, long-lasting, and intense. It is the major social activity after the young begin to loosen their bonds with their mothers and seek wider social relationships. The influence of such a pervasive activity on social development must be great. "If a large troop remains in one spot very long, as many as four different play groups may form. Ranging from the cluster of black infants inside the circle formed by their mothers to a boisterous group of older juveniles at the troop's edge, these four play groups represent the infants born during the four preceding annual birth peaks" (DeVore, 1963).

The social structure of the hamadryas baboons, though based on similar principles, differs in practice. The one-male units are the stable social groups within a more fluid set of social relationships between the units.

The nature of the play relationships reflects these differences. Because females are being integrated into the one-male units at an early age, through social bonds with their new male unit leader that are not unlike the earlier mother–infant bond, their freedom to join the play groups is seriously restricted. Unlike the savannah baboon female, which plays until she is about three years of age, the hamadryas female older than one year plays only if the play group is near a subadult or single adult male (Kummer, 1968a). The males, on the other hand, turn more and more to the peer group as the focus of their social activities. They often sleep together rather than with their mother's (and presumably their father's) one-male unit. This may be as much a factor of the weak mother–infant ties among hamadryas baboons as it is an indication of the strength of the peer-group bond. But the result of these combined factors is that male hamadryas have a constant opportunity to form a social bond with their age-mates even though each will, in later life, form a one-male unit that focuses most of the social relationships within the unit. But it is the male's social interaction with other males that binds these units together, and the bonds formed in play may well be crucial to the herd's stability.

The difference between male and female behavior is reinforced by the nature of play engaged in by the two sexes. Although the hamadryas baboons represent the utmost in differentiation, the male juvenile primate in many species plays more often and more roughly than the female. Moreover, in most species the male reaches social maturity at nearly twice the age of the female. Hamadryas females begin some aspects of adulthood as soon as they are taken into a harem between the ages of one to two years, whereas the male ordinarily does not become a unit leader until he is seven to ten years of age. The result is that the young male has several more years in which to play. Females generally avoid strong contact play such as wrestling.

Wrestling play, so common among social mammals, not only stimulates the fighting behavior of adults but also allows close contact with other individuals. During wrestling, the tactile stimulation is intense, and the advantages of contact between individuals in cementing social bonds are emphasized. It is quite possible that the bonds thus formed between aggressive males need a greater amount of tactile contact to counteract the higher levels of aggression.

Patas   The very different adaptation to savannah dwelling by patas monkeys is also manifested in their play. Since the social group is small, there are no age grades, and infants and juveniles of all ages play together, limited only by the control of the females when their small infants squeal in pain during the rougher play. In fact, female patas monkeys occasionally join in play with the young, but the adult males almost never do. The relatively peripheral position of the adult male in patas society may have developed, at least in part, because of the lack of male–infant social play (or vice versa).

Macaques: Bonnets, Japanese, and Rhesus   Bonnet macaque troops, especially the larger ones, resemble baboon troops in many respects. Play groups are age-graded to some extent but often consist of all ages of young and occasionally an adult male or two. The higher-intensity wrestling play separates the younger males and all of the females from the play group, which then form a play group of their own nearby. In this species, adult females never play with infants and juveniles, but adult males play not only with the younger animals but even with one another during the birth season. Their play is a mild form of wrestling characterized by hand grappling and mouthing. During the mating season, either the rising tension level in the troop prevents play between adult males or the free time otherwise spent in play is spent searching for sexually receptive females.

Thus bonnet macaque play forms bonds between the young that cut across mother-lineages somewhat like that of savannah baboons. Since growing males have a chance to play with adult males, they form bonds with them (and thus with the adult male association) before coming into aggressive conflict as they approach adulthood. This adult male–young male play probably enables the bonnet macaque male to remain an integrated member of his society throughout life and to avoid the peripheral or isolated period of late adolescence and early adulthood of some other species of macaques. The play of adult males during the birth season serves to strengthen the social bonds within the adult male association.

Japanese macaque infants and juveniles also play in peer groups that are drawn from a number of mother-lineages. Even in the Minoo-B troop, the infants and juveniles play across kin lines nearly as often as they do along kin lines (Yamada, 1963). Lacking an adult male association to focus social relationships, this troop is probably held together by the play bonds. The play groups of the large Takasakiyama troop are clearly formed along age-grade lines (Itani, 1954).

Infant rhesus macaques play much like the other macaques and baboons, with the same tendency to separate into age-graded play groups. Adult male play is apparently confined to occasional contact with juveniles and is an indication of their relationships with other males. Chance's (1956) report that the lowest-ranking of three adult males changed roles during the mating season from a fully integrated member of the adult male association to virtually a juvenile male who took part in juvenile play shows that although adult activities usually replace play, play fills the gap when adult outlets are unavailable.

Female rhesus in the laboratory (Harlow and Harlow, 1965) and bonnet macaques in the wild withdraw from rough-and-tumble wrestling. Both species maintain strong ties with their mothers and, unlike those of the males, the bonds formed by play are secondary. Males, on the other hand, must secure their position in society by establishing relationships with other males, partly through play, that are particularly important in troop organization.

Langurs The play patterns of Hanuman langurs resemble those of baboons and macaques, particularly in the large troop at Kaukori Village in north India (Jay, 1963, 1965). Infants and juveniles of the same size tend to segregate into distinct play groups, and young female langurs tend to avoid the rougher play of the young males; thus an early sex differentiation in behavior is established.

Among Hanuman langurs, the adult female controls the play group rather than the adult male, who remains aloof from the young until they are ten months old and begin to establish ". . . a relationship with adult males involving a series of special gestures and vocalizations" (Jay, 1963). In the weeks and months that follow, this relationship develops but does not seem to have the elements of play.

Characteristically the smaller Dharwar (central India) troops of Hanuman langurs do not have adult male associations; the single leader male in each troop sometimes plays with juveniles.

Chimpanzees It is becoming clear from Nishida's work on chimpanzees (1968) that the chimpanzee group is not an infinitely permeable association. Since the group does have boundaries, social bonds can be developed among all members of the group. Since the group is usually broken up into a series of independent subgroups that form and re-form in many different configurations, long-lasting social bonds must somehow be established if the group is to be an effective unit at any given time. As a matter of fact, most of the social bonds are developed and reinforced by play.

Several authorities have reported bands of chimpanzee mothers with their infants as one of the four general kinds of subgroupings in chimpanzee society (Kortlandt, 1962; Reynolds, 1965; Nishida, 1968). The fairly regular association of female chimpanzees and their offspring gives the young chimpanzees an opportunity to play with all the young of the total group. Moreover, the mixed bands, another of the four kinds of subgroupings, bring infants into contact with adult males.

Chimpanzee mothers often play with their own young but rarely with other infants (Van Lawick-Goodall, 1967), and adult males play with juveniles and infants. Thus among chimpanzees play bonds are formed within age groups as among baboons, between mother and offspring, and between young and adult males.

Gibbons The social group of gibbons is minimal, a mated pair and their offspring, and is not embedded in the larger social matrix of friendly relationships as in the hamadryas herd. The young gibbon, therefore, has

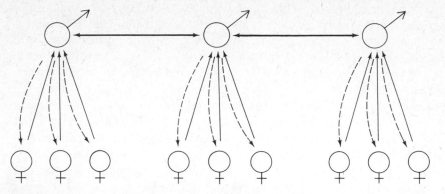

FIGURE 22 The three major bonds in hamadryas baboon society: Within the one-male unit, the bond between male and female is established by the male by pseudomothering and continued partially by sex (- - ➔). The female helps maintain the bond by grooming the male (——➔), and the relationship between one-male units is maintained by the male–male bond, which is probably established by an early play relationship (◄——➔).

a decidedly limited set of possible playmates, its siblings and its parents. Any contacts with neighboring groups are antagonistic and thus eliminate the possibility of bonds between the young of different groups. Adult males occasionally play with the young but most play is clearly between infants and juveniles (Carpenter, 1940).

Discussion

Much more research needs to be done before the role of play in the formation of social bonds can be stated with certainty. All play that occurs regularly over a fairly long period of time should help to form social bonds if contact is the primary factor in primate social bonding. However, different societies need different social bonds and the play groups of each should reflect these needs. The age-graded play groups of baboon troops, for example, contrast with the mixed-age playing in gibbon pairs. To clarify the nature of play bonds among different societies, field and laboratory workers must pay much more attention to the question of who plays with whom.

Those societies—particularly baboons, macaques, and probably vervets —that have evolved a high level of aggression as protection against predators require strong mechanisms for developing and reinforcing social bonds. As one of these mechanisms, play is likely to be organized to reinforce these bonds. Gorillas show little aggression and little play. But play is not the only bonding behavior nor is it used in all important relationships. Although in many primate societies the male uses play more than the female, the latter uses grooming to cement bonds she forms.

In the one-male unit of hamadryas baboons, grooming is the socially reinforcing activity that binds adult females to their males; on the other hand, the male's pseudomothering relationship initially establishes the bond that is reinforced by the later sexual relationship. Since among the one-male units the bonds are between the males themselves, with almost no contact between the females of separate units (and, in fact, with little bonding between the females of a unit), the more characteristically male activity, play, is used to establish an early and persistent bond between males.

Play relationships are probably more important for savannah baboons than for hamadryas since the bonds formed during youth establish not only a major bond cutting across mother-lineages but also dominance relationships before adult fighting occurs between age-mates (see below). Although the two-year birth interval probably did not evolve among baboons to establish the strongly age-graded play groups, nevertheless the birth interval does make play a more effective horizontal binding force precisely because it eliminates sibling playmates.

The peculiar thing about bonnet macaque society, reflected in the Somanathapur troop, is that both grooming and play are so strongly emphasized as binding forces. Whereas hamadryas and savannah baboons use grooming for one bond and play for another, bonnet macaque males groom a lot and play a lot. Further study may show that the more common small-sized forest troops of bonnet macaques use binding activities more sparingly than large social groups precisely because they do not face the special problems of the latter groups.

SOCIAL PLAY AS PREPARATION FOR ADULT LIFE

One of the common functions of play is to prepare for adult activities. Three major areas of adult activities can be isolated for this purpose: (1) those that are the province of females, or *maternal behavior;* (2) those that are generally the province of males, or *dominance activities;* and (3) those that involve both sexes, or *copulation.*

Maternal behavior can be learned through a kind of "playing house." The passing of infant Hanuman langurs among juvenile females as well as among adult females other than the mother allows the young female to gain experience in handling infants. Reynolds (1965) reports similar behavior for chimpanzees. Such activity on the part of the young females is similar to the male play of wrestling and chasing. It is not directed to immediate ends since the infant is not nursed and the real mother retrieves it at the first hint of danger.

By play-fighting, the young gain the skills needed to maintain their position of dominance in later life; hence it is usually the males that play-fight. In addition, they can learn by the play experience which ones can dominate the others and thus can reduce later fighting for position.

Finally, the adult heterosexual relationship requires learning, and that learning occurs optimally when the individual does not yet have a high level of sexual arousal. The play group is the ideal setting for this learning experience. Infants and juveniles mount each other and give pelvic thrusts, experimenting until they have achieved the proper positioning for copulation. In the normal social situation, an adult model is always available as an example. Laboratory studies by Mason (1963) and Harlow (1965) indicate that rhesus monkeys must learn the proper heterosexual relationship with other infants before they are twelve months old, or the patterns fail to develop properly.

Chapter 10
Sex and Society

For primates, sex is more than simply an act of reproduction. Man is not the only primate to use sex as an important instrument of social relationships, even as a binding force in society. This is possible because of the relatively high level of cortical control of sexual responses among primates. Male copulatory behavior triggered by the sight, sound, or smell of an estrous female and not modified by learned behavior patterns would be highly disruptive in a complex and permanent social organization that includes numerous males and females. But the ability to control sexual impulses through learned behavior allows various patterns to develop that can serve ends other than simple reproduction.

Recent laboratory studies have shown that primates cortically control their sexual responses. One such study, *Sexual Behavior in the Rhesus Monkey* (Harlow, 1965), suggests the learning necessary for appropriate sexual behavior in primates. During these studies, Harlow and his associates raised rhesus monkeys in isolation, with surrogate mothers, and in various degrees of contact with their own mothers and peers. They found that females raised from infancy without social contact could not posture properly for copulation; only after a long-term program of being placed with experienced males were four of these females impregnated:

one had been raised in a hardware-cloth cage; three had been raised on cloth surrogate mothers. Not only did their copulatory behavior remain undeveloped in the asocial setting but their whole reproductive capacity was also impaired.

Mason, in his work on social deprivation (1960, 1961, 1963, 1965), showed that similar problems face the male rhesus macaque deprived of the opportunity to learn sexual behavior in a social setting. For example, he compared rhesus males raised in the laboratory with those raised in the wild and found that the former rarely mounted in the appropriate position for copulation, whereas the free-ranging animals regularly mounted in the dorsoventral position and gave more pelvic thrusts than the cage-reared animals; moreover, their sexual interactions with the females lasted longer. Both sexes obviously need to learn appropriate sexual behavior in the social setting if their reproductive success is not to be markedly reduced.

The wild primate grows up in a social setting where it can observe adult copulatory behavior. In fact, the young commonly perform some sort of harassment during adult copulations. The copulating female's offspring, which is still in fairly close contact with its mother, comes close and threatens the male; moreover, peers of the young threatener are likely to join it. In addition, infants and juveniles often play at mounting and giving pelvic thrusts and thus learn normal copulatory posture long before they engage in actual copulation. Because the social contacts in a group of wild primates are regular and established, the animals do not have to establish a network of social relationships before they can copulate. On the other hand, when a caged rhesus female in estrus is taken from isolation and thrust in with an adult male, she must establish and develop a social relationship before she can copulate.

SEXUAL CYCLES

One of the common misconceptions about primates is that through most of the sexual cycle they have a humanoid sexuality characterized by female receptivity. Evidence over the last thirty years clearly shows that human patterns of sexual behavior have adapted to a particular kind of sociality from their common base of primate sexuality and that they differ rather remarkably from those of other primates and mammals. Nevertheless, the sexual cycles of man and the other primates share some similarities.

MENSTRUATION AND ESTRUS

The cycle that includes menstruation and estrus is common to the Old World monkeys, apes, and man. Some of the New World monkeys also menstruate, but prosimians do not, though they do have some bleeding into the uterine cavity as the corpus luteum disappears. True estrus, on the other hand, is common among nonhuman primates but appears vestigially in the human female, who often retains body fluids before menstruation in a manner similar to the estrous swelling of apes and monkeys.

In most higher primates, the estrous (and menstrual) cycle varies from about twenty to seventy days, with an average of around thirty days, and in many species is characterized by numerous physiological changes that are obvious either visually or olfactorily. The most common change among many species of macaques, baboons, and chimpanzees is a tendency to retain body fluids, which in extreme cases causes marked swelling in the perineal region, sometimes extending down the legs and along the dorsal surface of the sacrum and the base of the tail. The swollen region is often a bright red, which adds to the visual impact of the signal.

These obvious visual signs of estrus and therefore of receptivity are generally presented directly to the male. Thus the female initiates copulation, but once the relationship is established, the male takes the lead. Baboons, macaques, and chimpanzees turn their hindquarters to the solicited male and stand with the swellings only a short distance from his face. In the gelada female, the bright red perineal sex skin, bordered with white blister-like cutaneous vesicles, is duplicated in a chest patch of the same color that shows the same cyclical change in color intensity (Wickler, 1967). Thus the gelada, which sits while she eats, uses the chest patch to inform the male of her receptiveness; as he prepares to copulate, his approach is reinforced by the buttocks patch. He thereby avoids the mistake of approaching and attempting to copulate with a nonreceptive female. In a group with aggressive males, such a mistake could easily result in an attack if the female screeched in complaint. Vervets, which lack sexual swelling, rely on a form of posturing to signal sexual availability to the males.

Outside the mating season, sexual swelling probably helps to reduce social tensions. Since the swelling itself, the visual cue that initiates copulation, occurs only during the mating season, its absence outside the mating season eliminates male interest in copulation in those species in which the females take the initiative.

Among female baboons and female rhesus macaques, which initiate the sexual contact, this relationship very often develops into a consort relationship lasting several hours or even a few days. The consort pair removes itself from other social activity and is generally left without interference from the other males. Since the initiating presentation by the female is directed to the consort, he subsequently takes the initiative and the female sexual signals tend to be superseded and obscured. Other males are less likely to try to interfere with the pair.

CYCLES WITHOUT STRONG SIGNALS

Lacking the obvious swelling and color changes seen in many other species of macaques, bonnet macaques show two quite different patterns of matings. Hartman (1938) noted the absence of the bright red sex skin and observed instead the inconspicuous dark purple coloration. A pale red on the edges of the labia majora ran in lateral streaks from the ischial callosities and the tail base. The lack of color change and sexual swelling makes it extremely difficult to determine whether a bonnet macaque female is in estrus or not.

To compensate, male bonnet macaques often approach a female, lift her tail (or if she is sitting, first raise her buttocks until she stands), then examine her genitalia with their fingers and put their faces close, either to smell or to take a close look, or both. Although this approach occurs throughout the year, generally without copulation, it is the major activity of males during the June-to-October mating season, when most copulations take place. It is also the first phase of the usual pattern of solicitation for copulation. If receptive, the female stands in a sexually presenting posture, her tail raised or held to the side and her hindquarters raised somewhat more than her forequarters. Thus received, the male mounts, makes short, inserting pelvic thrusts followed by longer copulatory pelvic thrusts, terminates with an ejaculatory pause, and finally dismounts. Copulations are almost always completed in a single mount; the male and female usually separate immediately afterward, though they sometimes groom each other briefly. With this pattern of copulation, no consort relationships are formed.

The second form of mating behavior among bonnet macaques is initiated by the female. This rare female solicitation is engaged in almost exclusively by the young adult female, which on first coming into the sexual cycle is not yet recognized by the males for her sexual readiness. Lacking the extensive sexual swellings of the rhesus female, the female bonnet macaque resorts to behavioral signals.

Approaching a male that is not actively searching for receptive females, the young female comes close to him, looks into his face without the intensity of threat, and touches his face again and again with her hand. In the early stages, the female backs away if the male reaches for her. She then walks past him, slowing down or pausing as she goes by, and then sits about six feet away from him. If he does not follow, she returns and passes him again, perhaps looking into his face and touching it with her hand. The female repeats this approach pattern numerous times, each time backing away or moving on as the male reaches for her. After ten to thirty minutes, the male suddenly begins to jaw rapidly (rapid jaw movements, with the lips drawn back to expose the teeth), moves quickly to the female, mounts, and copulates without stopping to examine her genitalia. The female often gives a copulatory squeal, accompanied by a grimace, and the male continues to jaw during copulation. Often, instead of separating, the two continue in a consort relationship for several hours.

Thus, the first pattern—male investigation of the female genitalia to determine by sight and smell their condition—leads to a short, copulatory relationship that leaves the female free to approach or be approached by other males. The second—the elaborate, sometimes lengthy ritual of approach and withdrawal by the female—leads to a consort relationship similar to that of rhesus macaques.

MALE SEXUAL CYCLES

There is evidence that the males of some primate species undergo sexual cycles. Sade (1964) found seasonal changes in the sexual skin and testes size of male rhesus macaques on Cayo Santiago. The climate in Puerto Rico is well defined, and the rhesus macaques of Cayo Santiago have a well-defined birth season. Mating is restricted not only because the females are not receptive outside the fall mating season but also because the males undergo physiological changes as well. During the mating season, the testes enlarge and the scrotum and surrounding skin redden. Conoway and Sade (1965) have shown that in the late summer the seminiferous tubules enlarge and spermatogenesis begins, reaching a maximum during the fall breeding season. Once the season is over, the seminiferous tubules regress and spermatogenesis ceases.

Although no other systematic studies of male sexual cycles in primates have been reported, there is good reason to expect that such cycles occur in some of the other species with discrete mating and birth seasons. Higher primates vary rather widely in the matter of birth seasonality. Some species, like the rhesus of Cayo Santiago, have discrete birth seasons, outside of which there are virtually no births. The bonnet macaques in South India show birth peaks during a few months, but some births are scattered throughout the year. At the opposite extreme, neither the howler monkeys in South America nor the mangabeys in Africa have a discrete birth season or show a marked peaking of births at any time of the year; neither species has marked sexual cycles.

Japanese macaques, like the rhesus, have discrete birth seasons, but the islands of Japan extend over ten degrees of latitude and, because of their mountainous nature, vary widely in local environment. Moreover, the climatic changes are not sharply defined. Under such circumstances, birth seasons vary somewhat from group to group. Kawai, Azuma, and Yoshiba (1967) examined differences in latitude, rainfall, and regional variations but did not come to any clear conclusions about what triggers copulations and thus births.

The savannah baboons combine a birth peak with a scattering of births throughout the year (Lancaster and Lee, 1965). The males probably do not have any marked cyclical sexual changes. Carpenter (1934) found howler infants throughout the year and during a single period of collecting found embryos at several stages of development.

PREGNANCY, BIRTH, ESTRUS, AND RECEPTIVITY

Species like the rhesus macaque obviously cannot have continuous receptivity and, therefore, continuous sexual activity between males and females. Females are receptive for only a few short months and even then their receptivity is restricted to the days in which they are actually in estrus. At the most, this amounts to ten to fifteen days a month (Carpenter, 1942), during the first and last days of which female receptivity and male interest wax and wane. An adult rhesus macaque that spends 10 percent of her time in active sexual participation is probably an exception. Baboon females, whose births are spaced over a longer period, eighteen months to two years instead of one year, have an even lower percentage of active sexual receptivity (Rowell, 1967a).

Once the female primate is pregnant, she faces four, six, or eight months of pregnancy and as many months of nursing her infant before she becomes receptive again. The exceptions to this pattern may be significant when we consider whether primate sexual activity is a disruptive or a binding force in primate society. One of the exceptions is the bonnet macaque, which does not interrupt her sexual activity when she becomes pregnant. The mating season extends from the beginning of the monsoon rains in late June to September, by which time most females are pregnant. But I have observed copulations, though at ever-decreasing frequency, into November, December, and January. The females involved in these copulations began giving birth in January and thus must have copulated two, three, and four months after they had become pregnant. Among bonnet macaques, female sexual activity is spread over at least six months, sometimes eight, although births are restricted to about three months. Similar copulations during pregnancy are reported for vervets (Struhsacker, 1967b) and for rhesus macaques (Altmann, 1962; Conoway and Koford, 1965; Kaufman, 1967). Chalmers (1968b) observed mangabey males in Uganda copulating with nonswollen females, an indication that their sexual activity is not restricted to the peak of estrus.

On the other hand, the sexual activity of hamadryas baboons shows how restricted a male primate's sexual activity can be even in the midst of a large social group. The adult male harem leader copulates only with his own females. If he has three adult females in his harem, he probably has access to an estrous female not more than ninety days in a two-year period, assuming that the female is impregnated by her third estrus and that each female comes into estrus while the others are anestrus and therefore not receptive.

RANK AND COPULATORY SUCCESS

Zuckerman (1932) and Carpenter (1942) were the first to suggest the relation between dominance rank and successful copulation among primates. The receptive female, regardless of her rank, is almost always the center of attention. Occasionally, however, a particular female is not desirable or attractive to the males despite her estrus status. For example, one female bonnet macaque was ignored through several estrous cycles even though she presented to the males regularly and had some sexual swelling. Her problem was not obvious to the observer unaware of the standards of attractiveness used by the bonnet macaque males, who clearly showed no interest in copulating with her. However, the males are the ones to be considered in this discussion of dominance and sexual success.

Carpenter (1942) found that estrous rhesus females tended to copulate with one to three males and that they copulated first with subordinate males during the early phases of estrus. At the peak of estrus, they consorted with dominant males, only to return to the subordinate males as their estrus waned. DeVore (1965) described a similar situation among the savannah baboons:

Juvenile and young adult males copulate with estrous females during the female's periods of partial tumescence, and no bond between the female and her consort is established. Dominant adult males copulate with a female only during her period of maximum tumescence, a period which coincides with ovulation. A dominant male usually stays with an estrous female for several days, copulating with her exclusively and other troop members make little or no attempt to interfere with the pair.

Both these examples show that in these species dominant animals have the best opportunity to reproduce. But the situation is not quite that simple. DeVore also points out that among savannah baboons male dominance, and thus access to ovulating females, depends more on the alliances previously formed than on individual dominance. Old males, which in their toothless condition cannot dominate younger and stronger males, can do so if they are allied to a dominant male or form an alliance that dominates other large males. Thus the males in the adult male association, not just the most dominant individual male, have access to the estrous females.

On the other hand, among the hamadryas baboons the younger males with recently formed harems tend to have more females in their harem groups. With age, the harem leader usually attains higher dominance and social leadership. but he also loses females and herds fewer of them in his later years. Actually, the oldest males often cease to lead harems and thereby renounce their opportunity to reproduce since harem leaders and adult males generally copulate only with their own females. The younger males, in turn, copulate with more females and therefore reproduce at a higher rate than the older and more dominant males. Thus among the hamadryas baboons, the relation beween dominance and reproductive prowess is weak since the younger males, which are sexually active, have not yet been selected for social dominance. The old savannah baboon male, on the other hand, if he has been fortunate or skilled enough to form the proper alliances, sometimes has notable reproductive success.

Evidence from other sources casts further doubt on the correlation between dominance and copulatory or reproductive success. Consort relationships between subordinate rhesus males and estrous females have been reported in wild Indian groups (Southwick, Beg, and Siddiqi, 1965).

PLATE 19   A male Formosan macaque (*Macaca cyclopsis*), tail raised in domi-nance, gives an open-mouth threat.

The semiterrestrial *Lemur catta* have dominance hierarchies similar to those of macaques, but the access of males to receptive females does not seem to be based on rank (Jolly, 1967). Jay (1965) noted the same lack of correlation between dominance and copulatory success in the semi-ground-dwelling Hanuman langur.

Groups in captivity or otherwise confined to restricted space typically force out some males or otherwise prevent their mating. Chance (1956), working with a group of rhesus macaques in the London Zoo, observed that during the mating period the lowest ranking of three adult males was forced to renounce his position as a member of the adult male association and to assume the role of a juvenile, excluded not only from mating but also from the center of the group. In the wild, the availability of space and the wide variety of foliage, underbrush, and local topography would probably have enabled him to participate in mating, even if somewhat furtively.

Since the rank of the individual is determined far more by that of its mother (Kawai, 1958; Kawamura, 1958; Gartlan, 1968) than by some still unknown genetically inherited capacity for dominance, the importance of the father's rank is slight. If characters of dominance are genetically inherited, they are strongly influenced, if not often entirely overshadowed, by learned behavior in the social situation. The physical characteristics of strength and canine tooth size undoubtedly affect an animal's dominance if it has to fight its way to the top, but many adult male primates of average size probably rise as much by strength of character (their ability to ally themselves with other males) as they do through physical strength.

SEX AS A DISRUPTIVE FORCE IN SOCIETY

That sex is a disruptive force in many situations has been documented. DeVore (1965) writes: "When [baboon] group tension is high, the sight of a copulating pair may precipitate threatening charges by other males in the troop. These are usually adult males, but juveniles sometimes charge a copulating pair." As a result, savannah baboon consort pairs tend to move to the edge of the troop and thus to avoid social contact with other troop members. While this withdrawal removes a potentially disruptive activity from the center of social interaction, it also partially severs the social ties between the copulating pair and other members of the society.

Since primate sexual behavior is a series of learned patterns, the potential for variation in the behavioral patterns of a single species is considerable. Only a few groups of the many species studied in recent years have been sampled; therefore we cannot be sure that we are dealing with species specific patterns of behavior. The variability depends in different degrees on the ratios of males to females, on the personalities of individuals, and on idiosyncratic patterns that have spread and become a "custom" not only within a single troop but also within a series of neighboring troops with which the troop has exchanged males. For example, Chalmers (1968b), reporting on groups of mangabeys in Uganda, observed that the aggression in one group was almost completely confined to adult males squabbling over estrous females and that such conflict was totally absent in another group. If studied in isolation, the first group might seem to be a clear case of the disruptive influence of sexual relations on society and one might be tempted to label the mangabeys as a highly aggressive species. However, the lack of sexual aggression in the second group indicates that not sex but some other demographic influences or learned patterns of behavior combine to disrupt society.

Another example of sex as a disruptive force in society is found in Baldwin's (1968) description of squirrel monkey society. Outside the mating season, the males are neutral social factors more or less isolated from the female core of society and from one another. But during the mating season, the males come together and establish a dominance relationship. Baldwin describes four adult males that during the mating season traveled in a body and frequently came together for penile-display dominance interactions. The mating of such captive squirrel monkeys is accompanied by some trouble for the males that do not act effectively in concert even though they have established a dominance relationship. Approaching the females, they are often chased away by the latter, who are far more socially integrated. An individual male approaching a receptive female is often heckled by other females and can achieve successful copulation only when the receptive female is fairly isolated. In fact, Baldwin's study would indicate that successful copulation depends a good deal on the male's being relatively quiet and not overly excited when he approaches the females; highly excited males, making dominance displays and a lot of noise, attract heckling females that disrupt the copulatory sequence. Besides being a major source of conflict, squirrel monkey sexuality is also an expression of their basic social structure.

SEX AS A BINDING FORCE

According to one view, to produce a permanent social bond between members of a society, sex must be a year-round activity. Therefore a definite mating season rules out sex as a social bond during the remainder of the year. However, it must be stressed that primates not only learn but remember; hence the seasonal reinforcement of a social bond, helped along by other kinds of social bonds, probably suffices to make seasonal sex an important factor in society. If play, which usually terminates as the individual approaches adulthood, is a bond-forming agent with persisting effects, then sex, with its yearly reactivation, must be one also. Certainly there is evidence that the broad range of male-female bonds do persist for long periods outside the mating season.

Primitive tree shrews exemplify how sex can be an important factor in the formation of troop-like societies. Sorenson and Conoway (1968) established a colony of twelve (seven male and five female) *Tupaia montana*, or mountain tree shrews, living together in a single room. These shrews in Borneo exist as discrete social units of eight to twelve animals with a troop-like organization. Other tree shrews, such as *T. chinensis* and *T. longipes,* have not been seen in such large groups in the wild. Sorenson and Conoway feel that the high level of group vocalization among mountain tree shrews partly accounts for the large natural social units; four or five members will often vocalize when a disturbance occurs. On the other hand, in a caged group of *T. minor* and *T. longipes,* including (unnaturally, it is supposed) several males, the warning call is given only by the dominant male.

More important for this discussion, among female *T. montana,* in captivity at least, estrus continues for a longer period than among other species. The *T. montana* female accepts several males, and her receptivity is advertised to other members of the group by an ejaculation call emitted by the copulating male. This sharing of the females by the males could be a means of binding together the adult male association. Among *T. chinensis* and *T. longipes,* only the dominant male in the cage copulates with the estrous female. *T. longipes* have a minimal social organization and the females have an extremely short estrous cycle.

In captivity, *T. montana* appear to have no seasonal breeding or seasonal births. Whether this is peculiar to this species or due to the caged animals' being fed fresh fruit every day while living in a stable, season-free environment is not clear. If they truly lack copulation and birth seasons, the binding effect could be more pronounced.

Hamadryas Baboons

Several features of hamadryas baboon sexuality tend to bind the individuals of the society together. First, the hamadryas female becomes sexually receptive about one year before she is capable of conceiving. At an early age, often when she is only one year old, she is drawn into a permanent consort relationship in a kind of embryonic harem and begins copulating with the male at about the age of two (Kummer, 1968a). Copulation strengthens the bond between the pair at this point; certainly if sex were only disruptive, there is little possibility that this early, nonreproductive copulation would occur.

Second, among adult male hamadryas baboons, the harem leader does not copulate with females outside his harem. Extra-harem sexual relations are restricted to harem females copulating with juveniles and young adult males that have not yet established a harem. Once the male becomes a harem leader, he restricts his sexual activity to his own harem even though none of his females are mature enough for copulation. Thus the sexual bond is very strong for the male, perhaps less so for the female.

Bonnet Macaques

The sexual activity of bonnet macaques differs from that of other known macaques. Not only do they lack the sexual swelling characteristic of estrous rhesus macaques and baboons, but also the principal mating pattern makes for fairly rapid copulation and does not result in any consort relationship. Subadult males, adult males, and even juvenile males copulate within sight of each other without threat or other agonistic behavior. I have observed a female copulating within four minutes with the gamma male, a high-ranking subadult male, and finally a juvenile male. During the same period, the gamma male copulated with a second female, and all remained within a few feet of one another during the series of copulations. Only when a male persists in his advances to an unwilling female does sex prove a disruptive social force.

As a result of the free, or relatively free, access of all males in a bonnet macaque society to receptive females, there probably is less pressure on growing males to leave the troop. Isolated bonnet macaque males are very rare, and subadult males are free to move through the central part of the troop without threat.

THE SOCIAL ROLE OF SEX

One cannot generalize about either the binding or disruptive force of sex in primate society; too much depends on particular circumstances. Sex among primates has been woven into the very fabric of society. It is the basis of society in a pair-bond relationship such as is seen in gibbon society. On the other hand, in a society where the ratio of males to females is strongly overbalanced in favor of the males, it can be highly disruptive. Zuckerman (1932) described a group of zoo hamadryas baboons in which the few females were torn apart by competitive males.

Under favorable circumstances, the sexual relationship—like the mother–infant bond, the peer play-group bond, and the adult male bond in an adult male association—is a strong, positive influence on the permanence and strength of the society.

Chapter 11
The Adaptive Nature of Primate Society

We have discussed primate society primarily in terms of its internal structure and the variety of organizations that are possible with the basic social bonds formed by primates. But the social group is not isolated from the rest of the ecology; it is a functioning part of the ecosystem. Although information about the ecological relationships of primates is scanty, it is still fruitful to consider the evidence for ecological pressures on primate societies and to see what responses those societies are able to make.

ADAPTATION OF PRIMATE SOCIETY TO DIFFERENT NICHES

At least 90 percent of living nonhuman primates are arboreal (Napier, 1962). The other 10 percent live on savannahs and steppes and in close association with human beings in towns, temples, and cultivated areas. In these latter areas, if trees are available, they are used as part of the local primate habitat.

Most of the primates live in tropical areas where there are no marked seasonal changes in temperature such as are found in the temperate zones.

However, a number of species, including macaques and some langurs, range into regions of snow and frost; hence temperature apparently is not *the* major limiting factor for primates.

The social structure of those species that live outside the tropical forest is most likely to be strongly affected by the environment. Groups living in the tropical rain forests are to some extent buffered against excessive variations in food supply and pressure from predators by the safety afforded by the trees. Primates that have come to the ground at least partially and that have moved away from the trees onto the open savannah grasslands have had to alter their social structure to provide a new form of protection from predation (Washburn, 1968). The literature abounds in examples of the problems faced by these terrestrial primates.

ARBOREAL VERSUS SEMIGROUND-DWELLING PRIMATES

Most of the macaques, several species of *Cercopithecus* and *Cercocebus,* the Hanuman langur, chimpanzees, and gorillas spend a large proportion of their time on or near the ground. In their description of the ecological differences between *Cercocebus albigena* and *C. torquatus,* Jones and Sabater Pi (1968) have stressed how each has had to adapt its social structure to its use of the same forested area.

The entirely arboreal *Cercocebus albigena,* whose group size ranges from nine to eleven animals, inhabits the tall, dense vegetation of primary and secondary forests in the Congo basin and neighboring regions. Each group usually contains several females but only a single large adult male, which does not seem to be central to the group organization since he moves only peripherally with the group. They inhabit only the middle and upper canopies of the forest and their home ranges are small. One group, for example, lives in an area of 100 by 400 meters. In moving about, this species is restricted by brushy gaps in the forest, which, together with a fairly constant supply of food, tend to limit their movements to the same range during all seasons.

Cercocebus torquatus is found mostly in the dense vegetation of inland swamp forests, river margins, and parts of the littoral plain. Like many species of macaques and vervets, this species is terrestrial as well as arboreal, although they seem to move into open ground less than the macaques and vervets. *C. torquatus* groups spend considerable time on the ground and, in season, on mangrove roots above the swamp waters. Their group size ranges from fourteen to twenty-three animals, with several adult males in the larger groups. This is a typical troop organization, except that the groups occasionally split into subgroups that later re-form into the larger organization. This flexible group organization probably enables them to forage in small groups when predators are not threatening or when they are in the arboreal or swampy parts of their habitat. Then, in the dry season when they spend more time on the ground, they merge into the troop organization, which is safer. *C. tor-*

quatus groups occupy much more extensive ranges in the gallery forest as well as in the primary and secondary forests adjacent to rivers and swamps than *C. albigena* and range through a larger area during the dry season than during the rainy season.

These two species, which occupy different levels in the forest canopy, have adjusted their social organization to the requirements of their vertically differentiated habitats. In a habitat that is relatively safe from predation, *C. albigena*'s prime consideration is efficient foraging for food. Because the members of the small one-male groups can move together and eat near one another without overtaxing the food source, they do not need to occupy a large area to sustain themselves. Neither do the social bonds in the society need to be elaborate, since the close contact between the few members of the group is sufficient to maintain the group. The peripheral position of the male seems to indicate a weak social bond between him and the other members of the group. The mother-lineage is probably the primary social unit.

The flexible social organization of *C. torquatus*, on the other hand, enables the groups to adjust to the more variable habitat they occupy. They face seasonal changes that affect the sympatric *C. albigena* very little and must adjust to the changing arboreal and terrestrial conditions as they move into and out of the trees. Its ability to subdivide enables the group to move into the arboreal habitat when necessary with the same feeding efficiency as *C. albigena*. But when *C. torquatus* moves to the ground, the larger group formed by the reconsolidation of the sub-groups provides the protection of several adult males. The social organization emphasizes the male–male bond of the adult male association sufficiently to tie together the mother-lineages into a larger, more complex group. As might be expected, among the *C. torquatus* groups, the signals to coordinate troop movement are more efficient than those of *C. albigena*.

Although other examples are not identical, the rather similar relationship between Hanuman langurs and bonnet macaques in south Indian forests supports the generalization. Both species are somewhat more ground-dwelling than their African counterparts above, but the Hanuman langurs, which spend more time in the trees, have smaller groups and home ranges than the more terrestrial macaques. In nearby evergreen forests, troops of the highly arboreal lion-tailed macaque (*Macaca silenus*) are smaller than those of sympatric bonnet macaques.

SAVANNAH VERSUS DESERT

Within the various major ecologies are a number of subdivisions occupied by particular species or species groups. Therefore, in considering the ground-dwelling primates, we find that it is a matter of degree rather than of absolute dichotomy between them and the exclusively arboreal forms. Thus, *Cercocebus torquatus* can fall into the category of barely terrestrial; vervets and bonnet macaques into moderately terrestrial;

Japanese macaques, heavily terrestrial; and baboons, predominantly ter-restrial. Some species, such as *Papio hamadryas,* occupy virtual desert areas and are therefore completely terrestrial. The patas monkey lives in the open savannah grasslands and the gelada baboon occupies the montane, seasonally arid grasslands. Crook and Aldrich-Blake (1968) distinguish between savannah baboons on the basis of habitat; the olive baboons *(Papio anubis)* occupy the savannah woodlands of east and cen-tral Africa, *Papio cynocephalus* inhabit the forest fringe areas of central Africa, and *P. ursinus* occupy the south African savannah.

Since these various ground-dwelling species have been studied more extensively than any of the arboreal forms, we possess some data on the adaptation of their social organization to the ecology. The hamadryas and the savannah baboon as well as the patas monkey differ markedly in their social organization, and yet they all occupy a terrestrial habitat. We cannot, with the present meager evidence, demonstrate conclusively the reasons for these differences, but we can formulate some hypotheses about them.

Hamadryas baboons live in a dessicated area with no forest cover to afford them protection or concentrated supplies of food; hence they must move out over the open ground for fairly great distances to forage on the sparse vegetation. Since predator pressure is apparently fairly low, the protection provided by large troops moving together over the open ground is not essential. If the hamadryas social organization was once based on troop organization (and this seems to have been a distinct possibility—Kummer, 1968b), the emphasis now is on the male–female bond, not on the mother–infant and male–male (adult male association) bonds. The harem units have become the most stable in the society; and the daily foraging, though it begins with troop-like groups, often breaks down into the harem unit elements. However, presumably the vestigial adult male association allows large numbers of harem units to come together at the few available sleeping cliffs at night. The very marginal conditions in which these baboons live have necessitated a greater organi-zational flexibility than is usual for baboons. Their emphasis on subgroup-ings has enabled them to subdivide the large social groups when neces-sary, but their retention of male–male bonds also allows them to fuse again. Social bonds between the females are at a minimum, and this reduces or eliminates conflicting social bonds that would make it difficult for fission and fusion to take place on a regular basis.

Savannah baboons living in fairly open country in central and east Africa must face the large African predators—leopards, cheetahs, and lions—and cannot move out in small foraging groups. Thus foraging efficiently has had to be subordinated to predator protection. Since co-operation between large, powerful, aggressive males for the protection of the group is a *sine qua non,* the classical troop organization is found among these baboons. The adult male association allows more coopera-tion among males than any other subhuman primate organization. The bond between males and females is not an individual bond but a more diffused generalized one that does not lead to the permanent pairing of a single baboon male and a single or several females. On the other hand, savannah baboons living in more wooded areas have the added protection of nearby trees and so need not rely so heavily on the adult male association. For example, Rowell (1966, 1967b) has reported higher levels of intragroup aggression among the open grassland troops than among the forest baboons, an indication of the greater necessity to fight predators on the open grassland. But in spite of the heavier aggression, the troop organization is strong.

Patas monkeys occupy a habitat similar to that of the savannah baboons in the same areas of the Sudan. But even in the face of similar pressures, their social organizations are different. Patas monkeys rely on the adult male to serve as a decoy to lead predators away from the rest of the group. As a result, they have adapted anatomically by becoming the only well-adapted running primate. This adaptation enables the male to expose himself to the predator to distract it from the other group members and still escape once the predator has been led away. A single adult male is sufficient for this purpose; hence cooperation between adult males, which usually leads to an adult male association, has not developed in patas society. In fact, the patas male may not even be strongly bound to the females and young, but we still have little evidence about the nature of the bond between them.

It is clear that more than one social organization can adequately meet the demands of a particular environment, such as the savannah. It has been highly fashionable to compare and contrast baboon behavior with the adaptations of Pliocene tool-using, ground-dwelling apes. But it should be remembered that the long separation of our lines of descent may have led our early ancestors to a very different solution from that developed by either the baboons or the.patas monkeys (see Jolly, 1970).

Another caution should be issued. The ecological adaptations described here have been presented as if they were species-specific. But substantial evidence indicates that even among individual species the ability to adjust to different habitats varies and that some species are better than others at adapting to a range of environments. Forest-dwelling macaques are probably the most successful, since many species have adapted to village, town, and even city dwelling. Rowell (1966) observed forest baboons, which move over a home range that is considerably smaller than that occupied by the same species on the open savannah. The forest baboons obviously have a better opportunity to escape from aggressive interaction by running behind bushes and into the brush and trees. Hence the observer has the distinct impression that there is less emphasis on aggression among forest baboons than among savannah-dwelling baboons of the same species. Among the latter, moving out on the open grassland without any escape cover, aggression seems to play a much more important part in their daily life and dominance appears to be a primary social force.

IMPOVERISHED ENVIRONMENT

Cycles of dry years following lush, wet ones bring about changing conditions to which the animals in most parts of the world must adjust if they are to survive. During some seasons, they are forced to live in relatively impoverished conditions. Several field studies on primates have indicated the importance, in such circumstances, of variability within a single species, since the animals must adjust successfully to a wide range of conditions. Rowell (1967) cites an extreme case described by Hall. A troop of baboons trapped on an island formed by the rising water of the Kariba Dam in Rhodesia were in poor condition and short of food. These distressed animals departed from their usual troop organization and foraged in small groups or even singly. Moreover, social interaction was almost nonexistent: Infants and juveniles rarely played and sexual behavior was not observed. Rowell concludes: "One is tempted to suggest that where there is not time for anything but foraging, interactions become so few that social organization disintegrates."

Similar adaptations, though not so extreme, were reported by Gartlan and Brain (1968) in their study of the ecology and social variability of *Cercopithecus aethiops* (vervets) and *Cercopithecus mitis* (blue monkeys). The former were studied (1) in Lolui Island in Lake Victoria, a rich and regenerating environment and (2) in the Chobi area north of Lake Victoria, which the authors describe as "an impoverished and deteriorating habitat." The two vervet populations showed striking differences in social behavior.

On Lolui Island, the vervet groups had stable troop organizations that moved as close-knit units through their home ranges. The bonds between the adult male association and the females of the group were strong. In contrast, the Chobi vervets showed a stronger mother–infant bond and the individual spacing of the other group members was far greater than that of the Lolui Island vervets. "Distances of 500 meters were once observed between members of Group I, and distances of from 200 to 300 meters were common." Despite the presence of many predators, the Chobi vervets concentrated on gathering food rather than maintaining the protective troop organization. In terms of social organization, the mother-lineages were not subordinated sufficiently for the troop organization, with its strong bonds between males and between males and females, to be dominant.

Differences are not always attributable to impoverished versus rich environments. Jay (1965) described a strong troop social organization for the Kaukori langurs near Lucknow in northern India, in which the langurs spent a large proportion of their time on the ground near cultivated fields, where forest cover is severely restricted. Among the langurs near Dharwar in central India, which live in a forested region, Sugiyama (1967) describes a different social organization consisting of several one-male groups in small territories and a large all-male group living in the same area. One is tempted to conclude that the troop organization was an adaptation to the more open areas and that the smaller one-male group plus all-male group organization was more suitable to the forested regions. However, in spite of being forest-dwelling monkeys, the Hanuman langurs observed by Jay in the Bastar District of central India did not follow the one-male group–all-male group pattern described by Sugiyama. It is quite possible that the less demanding forest environment allows a freer range of social organization, much as relaxed selective pressures allow wider genetic variation in a breeding population.

In any event, many single species of primates are quite capable of adjusting to a variety of environments. Some are undoubtedly more flexible than others. The widespread and diverse species of macaques are certainly among the most flexible of the subhuman primates, but they are still remote from the extreme flexibility displayed by man with his use of culture as an adaptive mechanism.

RECIPROCAL ECOLOGICAL RELATIONSHIPS

The ecological relationships between plants and animals are never one-sided. Like all other creatures, monkeys must draw sustenance, shelter, and the other necessities of life from their surroundings. But they also contribute to the ecology and can alter it markedly under certain conditions. For example, one thinks of monkeys living in the forest and depending on it for survival, but rarely is the forest described as dependent on the monkeys for some of its characteristics or even for its survival.

The spread of forest plant species into neighboring or even distant areas is well known to be due, in part, to birds carrying seeds in their feces. Recent evidence indicates that monkeys likewise spread plant forms through their feeding habits and preferences. Gartlan and Brain (1968), in their study of the vervets of Lolui Island, described the role of the monkeys in the dispersal of fruits and seeds. After carefully examining the species of the thickets, they concluded that only one species had not been so dispersed and that the frugivorous vervets were the chief agents of dispersal. In their own way, these animals were at once engaging in a program of "reforestation" and expanding their own food resources. Moreover, the site for this reforestation was highly advantageous. New thickets were often established among rocks, where young seedlings have a better chance to grow than in the open grasslands, where fire, poor drainage, and impoverished soil threaten their survival. Furthermore, as the authors observed: "The monkeys, in their forays into the grasslands use the termitaria and rocks as resting places where they do much grooming, and as lookout spots over the tall grass; much urination and defecation occur here, and the high seed content of the feces . . . is left in an environment in which it is most likely to succeed when it germinates."

Since the monkeys spread the seeds of species palatable to them, the increase in monkey population tends to develop selectively and proportionally to the forest species they eat most. Certainly not only monkeys spread their favorite sources of vegetable food. Other faunal elements carrying plant seeds in their feces also contribute to the spread of plants and reforestation. But in a fairly limited faunal assemblage in which monkeys predominate, they are undoubtedly the primary means of spreading floral species; and the resulting ecological setting is markedly tailored to the monkeys' own food needs.

Oppenheimer and Lang (1969) found that the white-faced capuchin (*Cebus capucinus*) of the Barro Colorado Island in the Panama Canal Zone has a marked effect on the growth pattern of the membrillo tree (*Gustavia superba*). In two tracts of forest, one heavily populated with monkeys and the other lightly populated, the form of the membrillo trees differs markedly. Those on Barro Colorado Island, where the monkey population is high, have trunks that are more crooked and significantly more branched than the membrillo trees on the mainland site, where monkeys are rare. Over the long term, such excess branching produces a greater number of sites available for flowers and fruits in later years, although the immediate result is a lower yield. To show how such a significant effect can be produced by these animals, the authors described their feeding pattern. "While eating ends of the branches a monkey removes the bark along with some of the leaves, which are concentrated at the distal end. One or more inches of merismatic tissue and soft wood are then removed from the tip of the branch and eaten. Sometimes the end of the branch, with all the leaves, is broken off during feeding and dropped to the ground." They suggested that this feeding pattern releases the lateral buds by "removing apical dominance." This kind of interdependence must be reckoned with in an analysis of the ecological relationships between monkeys and the other aspects of their environment.

However, the results of environmental interdependence are not always favorable for monkeys. For example, because fever trees provide a major source of food as well as of escape for vervets throughout the year, they are an extremely important element in their habitat. Many of the herbs, grasses, and insects are only of seasonal importance, since they are abundant only after the rainy seasons. Other sources of food are exploited only occasionally. Eggs, for example, are apparently eaten whole; yellow-necked spur-fowl chicks provide occasional fare for the vervets, which kill them by biting off their heads.

The once ubiquitous and important fever tree is being seriously threatened in the Amboseli Reserve by the elephant population, which has been growing rapidly in recent years and has been felling fever trees at a far more rapid rate than they can be replaced naturally. The vervets, however, help to maintain and even spread fever trees. Hence the ecological relationship between the elephant population and the vervet population is a close one. But an overpopulation of elephants can drastically reduce the number of fever trees and thus in the long run also reduce the population of vervets.

Not all joint use of a food source is so competitive. In the forest, one of the methods to locate monkeys is to watch for deer eating under a tree. The monkeys usually pluck a fruit and take only one bite from it before dropping it to the forest floor. The deer eat the dropped fruit, which they otherwise would have little opportunity to get. In their movements, which range over wider areas than those of the monkeys, they also spread the seeds of the monkey's choice of food.

HOMINIDS AND NONHOMINIDS

It is axiomatic that primate society can adapt to whatever conditions prevail where the primates are found. Let us now turn briefly to human society to examine how it differs from that of subhuman primates and how these differences derive from man's adaptation to the ecology in which he lives.

Most of this presentation of primate behavior has emphasized the structural elements of primate society. Looking at the structures of non-human primates and human society, we must be struck by their great similarity. Human society has some added features and refinements that enhance man's flexibility in adjusting to new conditions, but the basic elements of human society can be recognized in subhuman primate society as well. The real difference between the two lies in language, for only with language can the categories of relatives be developed beyond the simple distinction between the members of one mother-lineage (the mother versus other close females) and also between members of one's own and other mother-lineages. Among males, either the adult male stands in a category by himself or there is a distinction between the members of the adult male association and nonmembers. These basic categories are recognized in the social interaction between monkeys and are probably conceptualized by the monkeys themselves. But any attempt to expand on this basic set of categories without language is probably too cumbersome and therefore impossible, and thus the limits of elaborating and expanding subhuman primate society are set by the lack of language.

When we compare the social bonds of human society with those of subhuman primates, we find that man combines all the primate social bonds in a single society. Man has the pair bond to form marriages, but the pair bond is strengthened by man's nonseasonal sexuality, a condition that is somewhat foreshadowed in bonnet macaque sexuality although without the pair bond. Man uses the mother-lineage or the mother–infant bond as the basis of many social relationships. The human mother–infant bond is apparently no more intense than the bond among nonhuman primates. In many ways, it is perhaps less so, since the human mother often leaves her infant with another member of the society without hurt to the child. This enables the infant to make contacts with other members of the society at an early age and to form social bonds with a variety of individuals, somewhat like the Hanuman langur infant when its mother passes it to other females.

The adult male association survives in primitive human societies in the male hunting band, the group of cooperating males needed in much of primitive human hunting. As human society became more complex, the adult male association diversified from the simple sexual and dominance distinction between central and peripheral males into a series of sub-groupings through the elaboration of the division of labor. The social bond formed by peers playing together in subhuman primate societies is found almost intact in human society. Among human beings, play is an extensive activity that continues as part of the learning process from childhood, through the growing years, and into adulthood or at least until sexual maturity. This extension of play into adulthood has likewise been observed among adult macaques.

Thus not only have the basic bonds of subhuman primate society survived in human society but in many cases human manipulation of them is similar to that found among subhuman primates. In addition, the combination of all of these social bonds in the larger human society has enabled man to exploit the bonding relationships far beyond any limits possible to nonhuman primates and to develop variations on the basic primate sociality. Moreover, with these bonds, together with the additional father–offspring bond (which is developed through a pair-bond relationship that is embedded in a strong matrix of the other bonds), man has produced a variety of kinship systems that are far more complex than any found among subhuman primates. The matrilineal and patri-lineal kinship systems found among so many of the preliterate human societies are highly adaptive systems whereby neighboring bands of human beings can be joined in alliance. These lineal and lateral extensions of kinship cannot take place without some relatively permanent social relationship to anchor the system. The pair-bond marriage is the most promising social relationship at hand since it includes both maternal and paternal relationships and thereby allows the kinship system to become egocentric instead of forcing it to rely on relationships between subgroups like the lineages. The primate mother-lineage, for example, whose recruitment is only through birth, could never have achieved the key breakthrough seen in human marriage, in which personal recruitment of an outside member has vastly expanded the social system.

All the societies developed by subhuman primates in varying levels of complexity have an outer limit and are thus basically closed systems. There is some exchange of members between such closed societies. Rowell (1966) reports that some male forest baboons shift from one troop to another, but in moving from one to become an integrated member of another, the individual must sever ties with his old society. Human beings have elaborated the social system not only by enlarging the society internally through various subgroupings like the mother-lineage and the adult male association, but also by combining larger neighboring societies through the pair-bonding system. In these large groupings, the individuals usually retain membership in the society of their birth and acquire it in the society of their marriage or pair bond. Very often a married woman who goes to live in her husband's group still retains her rights, duties, and privileges in the society in which she was born and gains other rights, duties, and privileges in the society into which she marries.

The lineal systems (patrilineal and matrilineal) and the nonunilineal systems of reckoning descent enable members to marry outside the system without forfeiting their ties with the group of their birth. Thus neighboring groups can develop a set of mutual rights and obligations that allow them to cooperate in endeavors that the individual group alone could not achieve. This is the kind of flexibility foreshadowed but never fully realized by the harem unit of hamadryas baboons and by the permeable subgroups of the chimpanzees.

Evolutionary Implications

Man's separation from other closely related primate lines of evolution must, according to the still scanty evidence, have happened a long time ago. If Leakey's *Kenyapithecus africanus* and *K. wickeri* are indeed man's ancestors, they indicate a line of descent already separate from the apes by late Oligocene times. Other evidence points to a separation in the Pliocene; hence we must look back on at least five million years of independent evolution in the hominid line. The forms most often compared with man are the baboons, since they occupy a similar ecological niche to that thought to have been occupied by the early hominids. Probably another five to ten million years, then, must be added to bring man's ancestry and that of the baboons to a common point. Fifteen to fifty million years of evolution allow for considerable differentiation. Thus any conclusions on the nature of basic primate behavior in the savannah, especially inherited capabilities and patterns, drawn from studies of baboons alone are necessarily shaky. The patas monkeys show another adaptation to very similar conditions, but it is unlikely that the patas organization would be the genesis of a successful hunting society in the hominid form. They would have had not only to overcome the problems of a dietary change but also to achieve a radical reorganization of their society.

By combining what we know about other species of primates, we may hope to eliminate some of the specializations developed by baboons during their separate evolution and thus to arrive at a common inheritance that we share with them and that must have been more obvious in earlier times, when the separation was more recent.

Assuming for the moment that man's hypothetical ancestry with the chimpanzee and gorilla holds any significance for his behavior (Reynolds, 1968), we can see in the flexible interactions between individuals that must have characterized early hominid social groups a reflection of the gorilla male tendency to rove from group to group during younger adult-hood and the permeability of the chimpanzee subgroups (with special emphasis on the male bands). In addition to this flexibility, the early hominids who moved out onto the open grasslands must have been weapon-users since they lacked large canines and would otherwise have had to remain close to the protection of the trees. But even tree-dwelling primates have large canines; thus it seems more likely that weapons must have replaced the large canines by Miocene times if the evidence of *Ramapithecus* (Simons and Pilbeam, 1965) is accepted. (For another view see Jolly, 1970). Early hominid adaptation to a terrestrial environ-ment must have been similar to that of the omnivorous–vegetarian baboons and chimpanzees; that is, it must have consisted mostly of vegetation, with occasional supplements of meat, eggs, and insects.

The drift away from a general primate and toward a specifically hominid adaptation probably came slowly as the climate cooled and dried during Miocene and Pliocene times. The slow shift to an omniv-orous–carnivorous diet may have taken 20 million years and begun well before the ancestral population divided into hominids and pongids. It is not necessary to postulate a sudden change from a predominantly vege-tarian to a predominantly carnivorous diet. Van Lawick-Goodall and Hall have both reported that chimpanzees and baboons respectively can eat quite a bit of meat, and Hall reports that some baboons apparently make meat a regular part of their diet. Once on the savannah, the early homi-nids would have found it necessary in lean years to rely more heavily on the herds of game animals available to them. Moreover, the slow changes in climate during the Miocene and Pliocene times were quite sufficient to put a slow but steady pressure on the savannah populations in the direction of an omnivorous–carnivorous diet.

The crucial breakthrough probably occurred when man became an efficient hunter instead of a casual hunter and reorganized his society around the hunting–gathering division of labor. At this time, stronger social bonds developed as bands of hunting males made forays far from the home base (an establishment necessitated by the need for a common meeting place). The male bond had to be strong enough to allow full cooperation of the hunting bands; as hunting grew in importance, the ability to form hordes of several local bands must have been particularly adaptive in hunting big game or carrying out complicated drives for smaller game. The rapid fluctuations of Pleistocene climate must have speeded the changes and made it necessary for hominids to adapt more efficiently. Efficient hunting first appeared in mid-Pleistocene times just after glaciations began. It is at this point that language became a neces-sity. The social system could not be expanded and manipulated into the human form unless categories could be assigned and named. Thus the expansion of the human brain must be closely tied to this revolution in hunting.

Field and laboratory studies have contributed to our understanding of the structure and evolution of subhuman primate societies and their adaptive nature. Depending on the adaptive needs of the group living in its particular ecological setting, the social bonds isolated and studied selectively in the laboratory have been shown by the field studies to be capable of manipulation into a variety of different organizations. The flexibility in adapting to different settings has been shown for several species of primates, most notably for macaques and baboons. Human beings have amplified this adaptive quality considerably through culture.

References

Allen, G. M.
   1939   A checklist of African mammals. *Bulletin of the Museum of Comparative Zoology,* vol. 83, pp. 1–763.
Altman, P. L., and D. S. Dittmer
   1964   *Biology data book.* Biophysics Laboratory, Aerospace Medical Research Laboratories, Wright Patterson Air Force Base, Ohio. AMRL-TR-64-100.
Altmann, S. A.
   1962   Social behavior of anthropoid primates: analysis of recent concepts. In *Roots of behavior: genetics, instinct, and socialization in animal behavior,* ed. Eugene L. Bliss, pp. 277–285. New York: Harper & Row.
Altum, J. B. T.
   1868   *Der Vogel und Sein Leben.* Munster: Niemann.
Andrew, R. J.
   1963   The origin and evolution of the calls and facial expressions of primates. *Behaviour,* vol. 20, pp. 1–109.
   1965   The origins of facial expression. *Scientific American,* no. 213, pp. 88–94.

1972   The information potentially available in mammal displays. In *Non-verbal communication,* ed. R. A. Hinde, pp. 179–206. Cambridge: Cambridge University Press.

Anthoney, T. R.
1968   The ontogeny of greeting, grooming, and sexual motor patterns. In Captive baboons (Superspecies *Papio cynocephalus*). *Behaviour,* vol. 31, parts 1–2, pp. 359–372, Leiden.

Asdell, S. A.
1964   Patterns of mammalian reproduction. Ithaca, N. Y.: Cornell University Press.

Baldwin, J. D.
1968   The social behavior of adult male squirrel monkeys (*Saimiri sciureus*) in a semi-natural environment. *Folia Primatologica,* vol. 9, nos. 3-4, pp. 281–314.

Bastian, J.
1968   Psychological perspectives. In *Animal Communication,* ed. T. A. Sebeok, pp. 572–591. Bloomington, Ind.: Indiana University Press.

Bates, B. C.
1970   Territorial behavior in primates: a review of recent field studies. *Primates, A Journal of Primatology,* vol. 11, no. 3, pp. 271–284.

Bernstein, I. S.
1965   Activity patterns in a *Cebus* monkey group. *Folia Primatologica,* vol. 3, pp. 211–224.
1968   The Lutong of Kuala Selangor. *Behaviour,* vol. 32, parts 1–3, pp. 1–16.

Bishop, A.
1962   Control of the hand in lower primates. *Annals of the New York Academy of Sciences,* no. 102, pp. 316–337.
1964   Use of the hand in lower primates. In *Evolutionary and genetic biology of primates,* vol. 2, ed. J. Buettner-Janusch, pp. 133–225. New York: Academic.

Blackwelder, R. E.
1967   *Taxonomy, a text and reference book.* New York: Wiley.

Bolwig, N.
1963   Bringing up a young monkey (*Erythrocebus patas*). *Behaviour,* vol. 21, pp. 300–330.

Booth, A. H.
1957   Observations on the natural history of the olive colobus monkey, *Procolobus verus* (van Beneden). *Proceedings of the Zoological Society of London,* vol. 129, part 3, pp. 421–430.
1958   Breeding of an immature wild baboon. *Journal of Mammalogy,* vol. 39, no. 3, p. 434 .

Booth, C.
1962   Some observations on the behavior of *Cercopithecus* monkeys.

Annals of the New York Academy of Sciences, vol. 102, pp. 477–486.

Bowden, D., P. Winter, and D. Ploog
1967 Pregnancy and delivery behavior in the squirrel monkey (*Saimiri sciureus*) and other primates. *Folia Primatologica,* vol. 5, pp. 1–42.

Buettner-Janusch, J.
1964 The breeding of galagos in captivity and some notes on their behavior. *Folia Primatologica,* vol. 2, no. 2, pp. 93–110.

1962 Use of the incisors by primates in grooming. *American Journal of Physical Anthropology,* vol. 20, pp. 129–132.

Bullard, Sir Edward
1969 The origins of the oceans. *Scientific American,* vol. 221, no. 3, pp. 66–75.

Burt, W. H.
1943 Territoriality and home range concepts as applied to mammals. *Journal of Mammalogy,* vol. 24, pp. 346–352.

Carpenter, C. R.
1934 *A field study of the behavior and social relations of howling monkeys* (Alouatta palliata). Comparative Psychology Monographs, vol. 10, no. 2, serial no. 48, pp. 1–168. Baltimore: Johns Hopkins Press.

1935 Behavior of red spider monkeys in Panama. *Journal of Mammalogy,* vol. 16, no. 3, pp. 171–180.

1937 An observational study of two captive mountain gorillas (*Gorilla beringei*). *Human Biology,* vol. 9, no. 2, pp. 175–196.

1940 *A field study in Siam of the behavior and social relations of the gibbon* (Hylobates lar). Comparative Psychology Monographs, vol. 16, no. 5, pp. 1–212. Baltimore: Johns Hopkins Press.

1941 The menstrual cycle and body temperature in two gibbons. In *Naturalistic behavior of nonhuman primates,* University Park, Penn.: Pennsylvania State University Press.

1942 Societies of monkeys and apes. In *Levels of integration in biological and social systems,* ed. R. Redfield, pp. 177–204. Biological Symposia, vol. 8. Lancaster, Penn.: The Jacques Cattell Press.

1964 *Naturalistic behavior of nonhuman primates.* University Park, Penn.: Pennsylvania State University Press.

1965 The howlers of Barro Colorado Island. In *Primate behavior: field studies of monkeys and apes,* ed. I. DeVore, pp. 250–291. New York: Holt, Rinehart and Winston.

Chalmers, N. R.
1968a Group composition, ecology and daily activities of free living mangabeys in Uganda. *Folia Primatologica,* vol. 8, nos. 3–4, pp. 247–262.

1968b The social behaviour of free living mangabeys in Uganda. *Folia Primatologica,* vol. 8, nos. 3-4, pp. 263–281.

Chance, M. R. A.
1956   Social structure of a colony of *Macaca mulatta. The British Journal of Animal Behaviour,* vol. 4, no. 1, pp. 1–13.
1961   The nature and special features of the instinctive social bond of primates. In *Social life of early man,* ed. S. L. Washburn, pp. 17–33. Chicago: Aldine.

Chance, M. R. A., and A. P. Mead
1953   Social behavior and primate evolution. In *Evolution, Symposia of the Society for Experimental Biology,* no. 7, pp. 395–439. New York: Academic.

Clark, Sir Wilfrid E. LeGros
1959   *Antecedents of man.* Chicago: Quadrangle.

Comfort, A.
1964   *The process of aging.* New York: Signet.

Conoway, C. H., and C. B. Koford
1965   Estrous cycles and mating behavior in a free-ranging band of rhesus monkeys. *Journal of Mammalogy,* vol. 45, pp. 577–588.

Conoway, C. H., and D. S. Sade
1965   The seasonal spermatogenic cycle in free ranging rhesus monkeys. *Folia Primatologica,* vol. 3, pp. 1–12.

Coolidge, H. J.
1936   Zoological results of the George Vanderbilt African Expedition of 1934. part IV: Notes on four gorillas from the Sanga River region. *Proceedings of the Academy of Natural Science,* vol. 88, 479–501.

Crook, J. H.
1966   Gelada baboon herd structure and movement, a comparative report. In *Play, exploration and territory in mammals,* ed. P. A. Jewell and C. Loizos, *Symposia of the Zoological Society of London,* no. 18, pp. 237–258.
1967   Evolutionary change in primate societies. *Science Journal,* vol. 3, no. 6, pp. 66–72.

Crook, J. H., and P. Aldrich-Blake
1968   Ecological and behavioral contrasts between sympatric ground dwelling primates in Ethiopia. *Folia Primatologica,* vol. 8, nos. 3–4, pp. 192–227.

Cullen, J. M.
1972   Some principles of animal communication. In *Non-verbal communication,* ed. R. A. Hinde, pp. 101–125. London: Cambridge University Press.

Darling, F.
1937   *A herd of red deer: a study in animal behavior.* London: Oxford University Press.

Devalois, R. L., and G.H. Jacobs
  1968   Primate color vision. *Science,* vol. 162, pp. 533–540.

DeVore, I.
  1963   Mother–infant relations in free-ranging baboons. In *Maternal behavior in mammals,* ed. H. L. Rheingold, pp. 305–335. New York: Wiley.
  1965   Male dominance and mating behavior in baboons. In *Sex and behavior,* ed. F. A. Beach, pp. 266–289. New York: Wiley.

DeVore, I., and K. R. L. Hall
  1965   Baboon ecology. In *Primate behavior: field studies of monkeys and apes,* ed. I. DeVore, pp. 20–52. New York: Holt, Rinehart and Winston.

DeVore, I., and S. L. Washburn
  1963   Baboon ecology and human evolution. In *African Ecology and Human Evolution,* ed. F. C. Howell and F. Bourliere. Viking Fund Publications in Anthropology, no. 36, pp. 335–367. New York: Wenner-Gren Foundation.

Eisenberg, J. F.
  1966   The social organization of mammals. In *Handbuch der Zoologie, Eine Naturgeschichte der Stamme des Tierreiches,* ed. W. Kukenthal and T. Krumbach. Berlin: Walter deGruyter.

Eisenberg, J. F., and R. E. Kuehn
  1966   The behavior of *Ateles geoffroyi* and related species. *Smithsonian Miscellaneous Collections,* vol. 151, no. 8, pub. no. 4683. Washington, D. C.: Smithsonian Institution.

Ellefson, J. O .
  1967   *A natural history of gibbons in the Malay Peninsula.* Ann Arbor, Mich.: University Microfilms.
  1968   Territorial behavior in the common white-handed gibbon, *Hylobates lar Linn.* In *Primates: studies in adaptation and variability,* ed. P. C. Jay, pp. 180–199. New York: Holt, Rinehart and Winston.

Etkin, W.
  1964a   Cooperation and competition in social behavior. In *Social behavior and organization among vertebrates,* ed. W. Etkin, pp. 1–34. Chicago: University of Chicago Press.
  1964b   Types of social organization in birds and mammals. In *Social behavior and organization among vertebrates,* ed. W. Etkin, pp. 256–297. Chicago: University of Chicago Press.

Fentress, J. C.
  1965   *Aspects of arousal and control in the behavior of voles.* Ph.D. Thesis, Cambridge, England.

Furuya, Y.
  1957   Grooming behavior in wild Japanese monkeys. In *Primates,* vol. 1, no. 1, pp. 47–68. Inuyama Yuen, Japan.

Gartlan, J. S.
   1968   Structure and function in primate society. In *Folia Primatologica,*
           vol. 8, pp. 89–120.
Gartlan, J. S., and C. K. Brain
   1968   Ecology and social variability in *Cercopithecus aethiops* and
           C. mitis. In *Primates: studies in adaptation and variability,* ed.
           P. C. Jay, pp. 253–292. New York: Holt, Rinehart and Winston.
Goodall, Jane
   1965   Chimpanzees of the Gombe Stream Reserve. In *Primate Be-
           havior: Field Studies of Monkeys and Apes,* ed. I. DeVore. New
           York: Holt, Rinehart and Winston.
Haddow, A. J.
   1952   Field and laboratory studies on an African monkey *Cercopi-
           thecus ascanius schmidti* Matschie. *Proceedings of the Zoological
           Society of London,* vol. 122, part 2, pp. 297–394.
Haldane, J. B. S.
   1955   Animal communication and the origin of human language. *Sci-
           ence Progress,* vol. 43, no. 171, pp. 385–401.
Hall, K. R. L.
   1960   Social vigilance behavior of the Chacma baboon, *Papio ursinus.*
           Behaviour, vol. 16, pp. 261–294.
   1962a   Numerical data, maintenance activities, and locomotion in the
           wild chacma baboon, *Papio ursinus. Proceedings of the Zoo-
           logical Society of London,* vol. 139, part 2, pp. 181–220.
   1962b   Sexual agonistic and derived social behavior patterns in the wild
           chacma baboon, *Papio ursinus. Proceedings of the Zoological
           Society of London,* vol. 139, part 2, pp. 283–327.
   1965   Experiment and quantification in the study of baboon behavior
           in the natural habitat. In *The baboon in medical research,* ed.
           H. Vagtborg, pp. 29–42. Austin: University of Texas Press.
   1967   Social interactions of the adult male and adult females of a
           patas monkey group, In *Social communication among primates,*
           ed. S. A. Altmann, pp. 261–280. Chicago: University of Chicago
           Press.
   1968a   Aggression in monkey and ape societies. In *Primates: studies in
           adaptation and variability,* ed. P. C. Jay, pp. 149–161. New
           York: Holt, Rinehart and Winston.
   1968b   Behaviour and ecology of the wild patas monkey, *Erythrocebus
           patas,* in Uganda. In *Primates: studies in adaptation and vari-
           ability,* ed. P. C. Jay, pp. 32–119. New York: Holt, Rinehart and
           Winston.
   1968c   Social organization of the Old-World monkeys and apes. In
           Primates: studies in adaptation and variability, ed. P. C. Jay,
           pp. 7–31. New York: Holt, Rinehart and Winston.
Hall, K. R. L., R. C. Boelkins, and M. J. Goswell
   1965   Behavior of the patas monkey, *Erythrocebus patas,* in captivity,

with notes on the natural habitat. *Folia Primatologica,* vol. 3, no. 1, pp. 22–49.

Hall, K. R. L., and I. DeVore
  1965 Baboon social behavior. In *Primate behavior: field studies of monkeys and apes,* ed. I. DeVore, pp. 53–110. New York: Holt, Rinehart and Winston.

Halliday, M. S.
  1966 Exploration and fear in the rat. In *Play, exploration and territory in mammals,* ed. P. A. Jewell and C. Loizos, *Symposia of the Zoological Society of London,* no. 18, pp. 45–60.

Hampton, J. K., Jr., S. H. Hampton, and B. T. Landwehr
  1966 Observations on a successful breeding colony of the marmoset, *Oedipomidas oedipus. Folia Primatologica,* vol. 4, pp. 265–287.

Harlow, H. F.
  1960 Primary affectional patterns in primates. *American Journal of Orthopsychiatry,* vol. 30, no. 4, pp. 676–684.

Harlow, H. F., and M. K. Harlow
  1965 The affectional systems. In *Behavior of nonhuman primates,* ed. A. M. Schrier, H. F. Harlow and F. Stollnitz, pp. 287–334. New York: Academic.

Harlow, H. F., M. K. Harlow, and E. W. Hansen
  1963 The maternal affectional system of rhesus monkeys. In *Maternal behavior in mammals,* ed. H. L. Rheingold, pp. 254–281. New York: Wiley.

Harlow, H. F., and R. F. Zimmerman
  1958 The development of affectional responses in infant monkeys. *Proceedings of the American Philosophical Society,* vol. 102, no. 5, pp. 501–509.

Hartman, C. G.
  1938 Some observations on the bonnet macaque. *Journal of Mammalogy,* vol. 19, no. 4, pp. 468–473.

Hediger, H. P.
  1961 The evolution of territorial behavior. In *The social life of early man,* ed. S. L. Washburn, pp. 34–57. Chicago: Aldine.

Hill, W. C. O.
  1942 The highland macaque of Ceylon. *Journal of the Bombay Natural History Society,* vol. 43, no. 38, p. 402.
  1953 Primates: comparative anatomy and taxonomy. I-Strepsirhini: a monograph. Edinburgh: University Press.
  1955 Primates: comparative anatomy and taxonomy. II-Haplorhini: Tarsioidea. Edinburg: University Press.
  1957 Primates: comparative anatomy and taxonomy. III-Pithecoidea: Platyrrhini (Families *Hapalidae* and *Callimiconidae*). Edinburgh: University Press.
  1960 Primates: comparative anatomy and taxonomy. IV-Cebidae: Part A. Edinburgh: University Press.

1962   Primates: comparative anatomy and taxonomy. V-Cebidae: Part B. Edinburgh: University Press.

1966   Primates: comparative anatomy and taxonomy, VI-Cercopithecoidea. Edinburgh: University Press.

Hinton, M. A. C., and R. C. Wroughton

1921   On the nomenclature of the South Indian long-tailed macaques. *Journal of the Bombay Natural History Society,* vol. 27, no. 4, p. 813.

Hooker, D.

1952   *The prenatal origin of behavior.* Lawrence, Kans.: University of Kansas Press.

1953   Early fetal behavior with a preliminary note on double simultaneous fetal stimulation. *Proceedings of the Association for Research in Nervous and Mental Disease,* vol. 33, pp. 98–113.

Howard, E.

1920   *Territory in bird life.* London: John Murray.

Humphries, D. A., and P. M. Driver

1967   Erratic display as a device against predators. *Science,* vol. 156, pp. 1767–1768.

Hutt, C.

1966   Exploration and play in children. In *Play, exploration and territory in mammals,* eds. P. A. Jewell and C. Loizos. Symposia of the Zoological Society of London, no. 18, pp. 61–82.

Imanishi, K.

1957   Social behavior in Japanese monkeys, *Macaca fuscata. Psychologia,* vol. 1, no. 1, pp. 47–54.

Itani, J.

1954   *The monkeys of Takasakiyama.* Tokyo: Kobunsa.

1959   Paternal care in the wild Japanese monkey, *Macaca fuscata fuscata. Primates,* vol. 2, no. 1, pp. 61–93.

1963   Vocal communication in the wild Japanese monkey. *Primates,* vol. 4, no. 2, pp. 11–66.

Itani, J., and A. Suzuki

1967   The social unit of chimpanzees. *Primates,* vol. 4, no. 2, pp. 355–382.

Jackson, G., and J. S. Gartlan

1965   The flora and fauna of Lolui Island, Lake Victoria. *Journal of Ecology,* vol. 53, pp. 573–597.

Jay, P.

1963   Mother–infant relations in langurs. In *Maternal behavior in mammals,* ed. H. L. Rheingold, pp. 282–304. New York: Wiley.

1965   The common langur of north India. In *Primate behavior: field studies of monkeys and apes,* ed. I. DeVore, pp. 197–249. New York: Holt, Rinehart and Winston.

Jewell, P. A.

1966   The concept of home range in mammals. In *Play, exploration and territory in mammals,* eds. P. A. Jewell and C. Loizos. Symposia of the Zoological Society of London, no. 18, pp. 85–109.

Jolly, A.
1966 *Lemur behavior.* Chicago: University of Chicago Press.
1967 Breeding synchrony in the wild *Lemur catta.* In *Social communication among primates,* ed. S. A. Altmann, pp. 3–14. Chicago: University of Chicago Press.

Jolly, C. J.
1970 Seed eaters: a new model of hominid differentiation based on a baboon analogy. *Man* (n.s.), vol. 5, no. 1, pp. 5–26.

Jones, C., and G. S. Pi
1968 Comparative ecology of *Cerocebus albigena* (Gray) and *Cercocebus torquatus* (Kerr) in Rio Muni, West Africa. *Folia Primatologica,* vol. 9, no. 2, pp. 99–113.

Kaufmann, J. H.
1962 Ecology and social behavior of the coati, *Nasua narica,* on Barro Colorado Island, Panama. University of California Publications in Zoology, no. 60, pp. 95–222.
1967 Social relations of adult males in a free-ranging band of rhesus monkeys. In *Social communication among primates,* ed. S. A. Altmann, pp. 73–98. Chicago: University of Chicago Press.

Kawai, M.
1958 On the system of social ranks in a natural troop of Japanese monkeys: (II) ranking order as observed among the monkeys on and near the test box. *Primates,* vol. 1–2, pp. 131–148.

Kawai, M., S. Azuma, and K. Yoshiba
1967 Ecological studies of reproduction in Japanese monkeys (*Macaca fuscata*) I: problems of the birth season. *Primates,* vol. 8, pp. 148–156.

Koford, C. B.
1963 Rank of mothers and sons in bands of rhesus monkeys. *Science,* vol. 141, pp. 356–357.

Kortlandt, A.
1962 Chimpanzees in the wild. *Scientific American,* vol. 206, no. 5, pp. 128–129.

Kummer, H.
1957 Soziales verhalten einer mantelpavian-gruppe. *Beiheft zur Schweizerischen Zeitschrift für Psychologie und ihre Anwendungen,* no. 33.
1967 Tripartite relations in hamadryas baboons. In *Social communication among primates,* ed. S. A. Altmann, pp. 63–72. Chicago: University of Chicago Press.
1968a *Social organization of hamadryas baboons.* Chicago: University of Chicago Press. Also published as *Bibliotheca Primatologica,* no. 6, eds. H. Hofer, A. H. Schultz and D. Starck. Basel and New York: S. Karger.

1968b Two variations in the social organization of baboons. In *Primates: studies in adaptation and variability*, ed. P. C. Jay, pp. 293–312. New York: Holt, Rinehart and Winston.

Kummer, H., and F. Kurt
1963 Social units of a free-living population of hamadryas baboons. *Folia Primatologica*, vol. 1, no. 1, pp. 4–19.

Lahiri, R. K., and C. H. Southwick
1966 Parental care in *Macaca sylvana*. *Folia Primatologica*, vol. 4, pp. 257–264.

Lancaster, J., and R. B. Lee
1965 The annual reproductive cycle in monkeys and apes. In *Primate behavior, field studies of monkeys and apes*, ed. I. DeVore, pp. 486–513. New York: Holt, Rinehart and Winston.

Leakey, L. S. B.
1967 An early Miocene member of Hominidae. *Nature*, vol. 213, no. 5072 (January 14), pp. 155–163.

Loizos, C.
1967 Play behaviour in higher primates: a review. In *Primate ethology*, ed. D. Morris, pp. 176–218. London: Weidenfeld and Nicolson.

Lumsden, W. H. R.
1951 The night-resting habits of monkeys in a small area on the edge of the Semliki Forest, Uganda. A study in relation to the epidemiology of Sylvan yellow fever. *Journal of Animal Ecology*, vol. 20, no. 1, pp. 11–30.

McKenna, M. C.
1963 New evidence against Tupaioid affinities of the mammalian family *Anagalidae*. *American Museum Novitiates*, no. 2158, pp. 1–16.

McNab, B. K.
1963 Bioenergetics and the determination of home range size. *The American Naturalist*, vol. 97, no. 894, pp. 133–140.

Marler, P.
1965 Communication in monkeys and apes. In *Primate behavior, field studies of monkeys and apes*, ed. I. DeVore, pp. 544–584. New York: Holt, Rinehart and Winston.

1968 Aggregation and dispersal: two functions in primate communication. In *Primates: studies in adaptation and variability*, ed. P. C. Jay, pp. 420–438. New York: Holt, Rinehart and Winston.

Marler, P., and W. J. Hamilton, III
1966 *Mechanisms of animal behavior*. New York: Wiley.

Malinowski, B.
1960 *A scientific theory of culture*. New York: Oxford University Press, Galaxy Books.

Maslow, A. H.
1936 The role of dominance in the social and sexual behavior of infrahuman primates: III. A theory of sexual behavior of infra-

human primates. *Journal of Genetic Psychology,* vol. 48, no. 2, pp. 310–338.

1940   Dominance-quality and social behavior in infrahuman primates. *Journal of Social Psychology,* vol. 2, no. 2, pp. 313–324.

Mason, W.

1960   The effects of social restriction on the behavior of rhesus monkeys: I. free social behavior. *Journal of Comparative and Physiological Psychology,* vol. 53, pp. 582–589.

1961   The effects of social restriction on the behavior of rhesus monkeys: II. tests of gregariosness. *Journal of Comparative and Physiological Psychology,* vol. 54, pp. 287–290.

1963   Social development of rhesus monkeys with restricted social experience. *Perceptual and Motor Skills,* vol. 16, pp. 263–270.

1965a   The social development of monkeys and apes. In *Primate behavior, field studies of monkeys and apes,* ed. I. DeVore, pp. 514–543. New York: Holt, Rinehart and Winston.

1965b   Determinants of social behavior in young chimpanzees. In *Behavior of nonhuman primates,* eds. A. M. Schrier, H. F. Harlow, and F. Stollnitz, pp. 335–364. New York: Academic.

1968   Use of space by *Callicebus* groups. In *Primates: Studies in adaptation and variability,* ed. P. C. Jay, pp. 200–216. New York: Holt, Rinehart and Winston.

Michael, R. P., J. Herbert, and J. Welegalla

1966   Ovarian hormones and grooming behaviour in the rhesus monkey (*Macaca mulatta*) under laboratory conditions. *The Journal of Endocrinology,* vol. 36, pp. 263–279.

Moffat, C. B.

1903   The spring rivalry of birds: Some views on the limit to multiplication. *Irish Naturalist,* vol. 12, pp. 152–156.

Montagna, W., and H. Machida

1966   The skin of primates. XXXII. The skin of the Philippine tarsier (*Tarsius syrichta*). *American Journal of Physical Anthropology,* vol. 25, pp. 71–84.

Napier, J.

1962   Monkeys and their habitats. *New Scientist,* vol. 15, no. 295, pp. 88–92.

Napier, J. R., and P. H. Napier

1967   *A handbook of living primates.* New York: Academic.

Neville, M. K.

1968   Ecology and activity of Himalayan foothill rhesus monkeys (*Macaca mulatta*). *Ecology,* vol. 49, pp. 110–122.

Nishida, T.

1968   The social group of wild chimpanzees in the Mahali Mountains. *Primates,* vol. 9, pp. 167–227.

Nissen, H. W.

1931   *A field study of the chimpanzee. Observations of chimpanzee behavior and environment in western French Guinea.* Comparative Psychology Monographs, vol. 8, no. 1, serial no. 36. Baltimore: Johns Hopkins Press.

Nolte, A.
  1955 Field observations on the daily routine and social behavior of common Indian monkeys, with special reference to the bonnet monkey (*Macaca radiata*). *Journal of the Bombay Natural History Society,* vol. 53, pp. 177–184.

Oppenheimer, J. R., and G. E. Lang
  1969 Cebus monkeys: effect on branching of *Gustavia* trees. *Science,* vol. 165, pp. 187–188.

Perkins, E., T. Arao, and H. Uno
  1968 The skin of primates. XXXVIII. The skin of the red uacari (*Cacajao rubicundus*). *American Journal of Physical Anthropology,* vol. 29, pp. 57–80.

Petter, J. J.
  1962 Ecological and behavioral studies of Madagascar lemurs in the field. *Annals of the New York Academy of Sciences,* vol. 102, Art. 2, *The Relatives of Man,* New York.
  1965 The lemurs of Madagascar. *Primate behavior: field studies of monkeys and apes,* ed. I. DeVore, pp. 292–319. New York: Holt, Rinehart and Winston.

Petter-Rousseaux, A.
  1964 Reproductive physiology and behavior of the Lemuroidea. In *Evolutionary and genetic biology of the primates,* vol. 2, ed. J. Buettner-Janusch, pp. 92–132. New York: Academic.

Piveteau, J.
  1957 *Traite de Paleontologie,* Tome VII, *Primates, Paleontologie Humaine.* Paris: Masson.

Ploog, D. W.
  1967 The behavior of squirrel monkeys (*Saïmiri sciureus*) as revealed by sociometry, bioacoustics, and brain stimulation. In *Social communication among primates,* ed. S. A. Altmann, pp. 149–184. Chicago: University of Chicago Press.

Ploog, D. W., J. Blitz, and F. Ploog
  1963 Studies on social and sexual behavior of the squirrel monkey (*Saimiri sciureus*). *Folia Primatologica,* vol. 1, no. 1, pp. 29–66.

Pocock, R. I.
  1931 Revision of the macaques. *Journal of the Bombay Natural History Society,* vol. 35.

Poirier, F. E.
  1967 The ecology and social behavior of the Nilgiri langur (*Presbytis johnii*) of South India. Ph.D. dissertation, University of Oregon.
  1968a The Nilgiri langur (*Presbytis johnii*) mother-infant dyad. (*Primates,* vol. 9, pp. 45–68.

Prechtl, H. F. R.
  1958 The directed head turning response and allied movements of the human baby. *Behaviour,* vol. 13, parts 3–4, pp. 212–242.

Reynolds, V.
    1965   Some behavioral comparisons between the chimpanzee and the mountain gorilla in the wild. *American Anthropologist,* vol. 67, pp. 691–706.
    1968   Kinship and the family in monkeys, apes and man. *Man* (n.s.), vol. 3, no. 2, pp. 209–223.
Rosenblum, L. A., I. C. Kaufman, and A. J. Stynes
    1964   Individual distance in two species of macaque. *Animal Behavior,* vol. 12, nos. 2–3, pp. 338–342.
Rowell, T. E.
    1963   Behaviour and female reproductive cycles of rhesus macaques. *Journal of Reproductive Fertility,* vol. 6, pp. 193–203.
    1966   Forest living baboons in Uganda. *Journal of Zoology,* vol. 149, part 3, pp. 344–364.
    1967a   Female reproductive cycles and the behavior of baboons and rhesus macaques. In *Social communication among primates,* ed. S. A. Altmann, pp. 91–96. Chicago: University of Chicago Press.
    1967b   Variability in the social organization of primates. In *Primate Ethology,* ed. D. Morris, pp. 219–235. London: Weidenfeld and Nicholson.
Rowell, T. E., and R. A. Hinde
    1963   Responses of rhesus monkeys to mildly stressful conditions. *Animal Behavior,* vol. XI, nos. 2–3, pp. 235–243.
Sade, D. S.
    1964   Seasonal cycle in size of tests of free-ranging *Macaca mulatta. Folia Primatologica,* vol. 2, no. 3, pp. 171–180.
    1965   Some aspects of parent-offspring and sibling relations in a group of rhesus monkeys, with a discussion of grooming. *American Journal of Physical Anthropology,* vol. 23, pp. 1–18.
    1967   Determinants of dominance in a group of free-ranging rhesus monkeys. In *Social communication among primates,* ed. S. A. Altmann, pp. 99–114. Chicago: University of Chicago Press.
Sanderson, I. T.
    1957   The monkey kingdom. Garden City, N. Y.: Hanover House.
Sauer, E. G. F.
    1967   Mother–infant relationship in galagos and the oral child-transport among primates. *Folia Primatologica,* vol. 7, no. 2, pp. 81–97.
Schaller, G. B.
    1963   *The mountain gorilla: ecology and behavior.* Chicago: University of Chicago Press.
    1965   The behavior of the mountain gorilla. In *Primate behavior, field studies of monkeys and apes,* ed. I. DeVore. New York: Holt, Rinehart and Winston.
Service, E.
    1962   *Primitive social organization.* New York: Random House.
Simonds, P. E.
    1965   The bonnet macaque in South India. In *Primate behavior: field*

studies of monkeys and apes, ed. I. DeVore, pp. 175–196. New York, Holt, Rinehart and Winston.

Simons, E., and D. Pilbeam
1965   Preliminary revision of the Dryopithecinae (Pongidae, Anthropoidea). *Folia Primatologica,* vol. 3, pp. 81–152.

Simpson, G. G.
1961   *Principles of animal taxonomy.* New York: Columbia University Press.

Sonek, A.
1966   Social behavior of a colony of woolly spider monkeys. M.A. thesis, University of Oregon.

Sorenson, M. W., and C. H. Conoway
1966   Observations on the social behavior of tree shrews in captivity. *Folia Primatologica,* vol. 4, pp. 124–145.
1968   The social and reproductive behavior of *Tupaia montana* in captivity. *Journal of Mammalogy,* vol. 49, no. 3, pp. 502–512.

Sparks, J.
1967   Allogrooming in primates: a review. In *Primate ethology,* ed. D. Morris, pp. 148–175. London: Weidenfeld and Nicolson.

Southwick, C. H., M. A. Beg, and M. R. Siddiqi
1965   Rhesus monkeys in North India. In *Primate behavior, field studies of monkey and apes,* ed. I. DeVore, pp. 111–159. New York: Holt, Rinehart and Winston.

Struhsaker, T. T.
1967a   Auditory communication among vervet monkeys *(Cercopithecus aethiops).* In *Social communication among primates,* ed. S. A. Altmann, pp. 280–324. Chicago: University of Chicago Press.
1967b   Ecology of vervet monkeys *(Cercopithecus aethiops)* in the Masai-Amboseli Game Reserve, Kenya. *Ecology,* vol. 48, pp. 891–904.

Sugiyama, Y.
1964   Group composition, population density, and some sociological observations of Hanuman langurs *(Presbytis entellus).* *Primates,* vol. 5, pp. 7–38.
1967   Social organization of Hanuman langurs. In *Social communication among primates,* ed. S. A. Altmann, pp. 221–236. Chicago: University of Chicago Press.

Sugiyama, Y., K. Yoshiba, and M. D. Parthasarathy
1965   Home range, mating season, male group, and intertroop relations in Hanuman langurs *(Presbytis entellus).* *Primates,* vol. 6, pp. 73–106.

Tappen, N. C.
1960   Problems of distribution and adaptation of the African monkeys. *Current Anthropology,* vol. 1, no. 2, pp. 91–120.

Tarling, D. H., and M. P. Tarling
1971   *Continental drift, a study of the earth's moving surface.* London: G. Bell and Sons.

Thorington, R. W., Jr.
  1968   Observations of the tamarin (*Saguinus midas*). *Folia Primatolog-ica*, vol. 9, no. 2, pp. 95–98.

Ulrich, W.
  1961   Zur Biologie und Soziologie der Colobusaffen (*Colobus guereza caudatus* Thomas 1885). In *Der Zoologische Garten* (NF), Bd. 25.6, pp. 305–368.

Van den Burghe, L.
  1959   Naissance d'un Gorille de Montagne à la Station de Zoologie Experimentale de Tshibati. *Folia Scientifica Africae Centralis*, vol. 4, pp. 81–83.

Van Lawick-Goodall, J.
  1967   Mother–offspring relationship in free-ranging chimpanzees. In *Primate ethology*, ed. D. Morris, pp. 287–346. London: Weiden-feld and Nicolson.
  1968   A preliminary report on expressive movements and communica-tion in the Gombe Stream chimpanzees. In *Primates: studies in adaptation and variability*, ed. P. C. Jay, pp. 313–374. New York: Holt, Rinehart and Winston.

Washburn, S. L.
  1968   Speculations on the problem of man's coming to the ground. In *Changing perspectives on man*, ed. B. Rothblatt, pp. 193–206. Chicago: University of Chicago Press.

Washburn, S. L., and D. A. Hamburg
  1965   The study of primate behavior. In *Primate behavior: field studies of monkeys and apes*, ed. I. DeVore, pp. 1–13. New York: Holt, Rinehart and Winston.
  1968   Aggressive behavior in Old World monkeys and apes. In *Pri-mates: studies in adaptation and variability*, ed. P. C. Jay, pp. 458–478. New York: Holt, Rinehart and Winston.

Washburn, S. L., P. C. Jay, and J. B. Lancaster
  1965   Field studies of Old World monkeys and apes. *Science*, vol. 150, no. 3703, pp. 1541–1547.

Washburn, S. L., and C. S. Lancaster
  1968   The evolution of hunting. In *Man the hunter*, eds. R. B. Lee and I. DeVore, pp. 293–303. Chicago: Aldine.

Webb-Peploe, C. G.
  1947   Field notes on the mammals of South Tinnevelly, South India, *Journal of the Bombay Natural History Society*, vol. 46, no. 4.

Wharton, C. H.
  1950a Notes on the Philippine tree shrew, *Urogale everetti* Thomas. *Journal of Mammalogy*, vol. 31, no. 3, pp. 352–354.
  1950b The tarsier in captivity. *Journal of Mammalogy*, vol. 31, no. 3, pp. 260–268.

Wickler, W.
  1967   Socio-sexual signals and their intra-specific imitation among

primates. In *Primate ethology,* ed. D. Morris, pp. 69–147. London: Weidenfeld and Nicolson.

Yamada, M.
   1963   A study of blood-relationship in the natural society of the Japanese macaque. *Primates,* vol. 4, no. 3, pp. 43–66.

Yerkes, R. M.
   1943   Chimpanzees: A laboratory colony. New Haven: Yale University Press.

Yoshiba, K.
   1968   Local and intertroop variability in ecology and social behavior of common Indian langurs. In *Primates: Studies in adaptation and variability,* ed. P. C. Jay, pp. 217–242. New York: Holt, Rinehart and Winston.

Zuckerman, S.
   1932   *The social life of monkeys and apes.* New York: Harcourt, Brace and Company.

Index of Authors

Taxonomic Index

Subject Index